CHANCE, LOGIC and INTUITION

An Introduction to the Counter-Intuitive Logic of Chance

CHANCE, LOGIC and INTUITION

An Introduction to the Counter-Intuitive Logic of Chance

Steven Tijms

NEW JERSEY • LONDON • SINGAPORE • BEIJING • SHANGHAI • HONG KONG • TAIPEI • CHENNAI • TOKYO

Published by

World Scientific Publishing Co. Pte. Ltd.

5 Toh Tuck Link, Singapore 596224

USA office: 27 Warren Street, Suite 401-402, Hackensack, NJ 07601

UK office: 57 Shelton Street, Covent Garden, London WC2H 9HE

Library of Congress Control Number: 2021003538

British Library Cataloguing-in-Publication Data
A catalogue record for this book is available from the British Library.

CHANCE, LOGIC AND INTUITION
An Introduction to the Counter-Intuitive Logic of Chance

ISBN 978-981-122-918-3 (hardcover)
ISBN 978-981-122-919-0 (ebook for institutions)
ISBN 978-981-122-920-6 (ebook for individuals)

For any available supplementary material, please visit
https://www.worldscientific.com/worldscibooks/10.1142/12069#t=suppl

Desk Editor: Soh Jing Wen

Printed in Singapore

"And thus, by combining the uncertainty of chance with the force of mathematical proof and by the reconciliation of two apparent opposites, she derives her name from both of them and rightfully assumes the wonderful name of Mathematics of chance!" — *Blaise Pascal*

Preface

Few people have won the Nobel Prize twice. Linus Pauling (1901–94) is one of them. The American chemist and peace activist was awarded the Nobel Prize in Chemistry in 1954 and followed this up with the Nobel Peace Prize seven years later. During the ceremony he told the audience that winning the Nobel Prize for the second time is not as exceptional as it seems. It's true, he admitted, that the chances of anyone winning a Nobel Prize for the first time are very small. But the chances of winning a second Nobel Prize are much better. After all, you only have to compete with others who have already won a Nobel Prize and who are still alive.

Pauling was joking, of course. But the anecdote illustrates how easily we are misled when it comes to probabilities. As Persi Diaconis, a well-known expert on probability, once put it: "Our brains are just not wired to do probability problems very well." When asked about the surprising solution to a famous probability problem known as the Monty Hall dilemma, he admitted that his intuitive response to similar problems had been wrong time and time again. When it comes to probabilities, our intuition is a poor guide.

Our intuition solves problems by taking shortcuts. Instead of analyzing the problem, it automatically translates it, so to speak, into a similar-looking problem with a familiar solution. It then solves the original problem by analogy. In psychology this intuitive form of thinking is qualified as *system 1 thinking*. Daniel Kahneman, another Nobel Prize winner and known for his work on the psychology of judgement, refers to it as *fast thinking*.

Fast thinking often works well enough, but not in the realm of chance, where familiar principles typically do not apply. Take, for example, the rolling of a die. In contrast to what many people intuitively expect, the probability of rolling a six doesn't increase with every unsuccessful roll of the die. Or, say, you bet on rolling at least one six in two consecutive rolls. In this case the overall chance is not the sum of the individual probabilities; it's less. When you're dealing with possibilities, one and one don't add up to two.

These misconceptions may betray a naïve understanding of probability, but also on a more advanced level of understanding our intuition tends to steer us in the wrong direction. In fact, we often find ourselves at a complete loss in the face of a probability problem. And yet the realm of chance is not entirely inaccessible, as the ancient Greek philosopher Aristotle once thought. The unpredictability of chance has a logic of its own, a logic that is governed by the rules of probability. So, can probability theory help us where our intuition fails? It certainly can, but in this regard a different kind of obstacle emerges.

Of all the branches of mathematics, probability theory is perhaps the closest to our daily lives. But if you open a standard book on probability, you may very well feel far removed from the reality of daily life. Mathematical symbols and formulas stare back at you with almost every turn of the page. Little is mentioned about the role of chance in our everyday lives. And even if you're not scared off by some abstract math, you may still be disappointed and close the book, never to open a book on probability again.

This book takes an entirely different approach. Instead of a formal presentation of probability, packed with formulas and definitions, the mathematics has been embedded into an everyday narrative context.

Part One tells the story of the "taming of chance". You will meet the pioneers of the mathematics of probability. Contrary to popular belief, these mathematicians did not live their lives secluded in a Platonic world of mathematical ideas. They kept a firm eye on the world around them. They also had their own dreams. Understanding how chance operates can free us from the notion that we are at the whim of a fickle goddess, or that we are fettered by the chains of fate. It can make us fair gamblers. It can even guide us through life's uncertainties.

These pioneering mathematicians were fascinating personalities, in no way inferior to their famous contemporaries in literature and philosophy. Some are still known, such as Galileo Galilei and Blaise Pascal, while others are almost forgotten, as, for instance, the colorful scholar Girolamo Cardano. Hopefully, this story about the birth of probability will bring them back to life.

What about the math in the book? You don't have to be a diehard mathematician to understand the mathematical parts of this founding story. The elementary skills of adding, subtracting, dividing, and multiplying numbers will do. By following in the footsteps of the pioneers of probability, you will learn, in a natural way, some basic concepts and simple rules of probability. At the same time, you will also learn to avoid the pitfalls that the founders of probability encountered in their path.

In Part Two we will explore the counter-intuitive logic of chance. Chance impacts everyday life in many different ways. The future is often uncertain and unpredictable, and our path is full of coincidences. You would expect us to have some understanding of the capricious ways of chance, but that is far from being the case. When we face chance in everyday life our intuition fools us time and again.

Some misconceptions have even been given their own names, such as the *fallacy of the transposed conditional* or, the most deceitful of all, the *gambler's fallacy*. They are called fallacies because their deceptive reasoning appears to be logical. It isn't. Despite seeming very convincing, they continually steer us in the wrong direction.

When we begin to truly understand chance it often appears paradoxical and hard to believe — not because it is really paradoxical, but because, at first glance, it runs completely counter to our intuition. However, it will stop looking like this once we have replaced our intuitive framing with an alternative way of looking.

Unfortunately, many of our intuitive misconceptions of the way chance operates are deeply rooted. They can only be eradicated by thinking slow. But equipped with a basic understanding of the rules of probability and a good dose of inventiveness, you will be able to rid yourself of these intuitive misconceptions. In the end, you'll gain a better understanding of the world we live in.

Wittgenstein once wrote that the purpose of his philosophy was "to show the fly the way out of the fly-bottle". I hope to achieve the same with regard to probability. Familiarity with the counter-intuitive logic of chance may prevent you from getting carried away by spurious kinds of wisdom. As Stephen Jay Gold once put it: "misunderstanding of probability may be the greatest of all general impediments to scientific literacy".

At the same time, this book can be read as the fascinating story of the birth of a new branch of mathematics. You may be surprised that the concept of probability, which is so familiar today, was introduced just three and a half centuries ago. You may also be surprised at how many applications this young branch of mathematics has acquired in daily life ever since.

But the main reason I wrote this book is that the reader will enjoy reading it.

Acknowledgments

Four years ago I embarked on a journey without knowing where it would take me. Luckily I had a compass with which to navigate and a safe haven when the weather was rough. Without the support of my parents, Els and Henk, I would have been hopelessly adrift. The journey resulted in "Toeval is altijd logisch" (Chance is a logical thing), a Dutch-language book which served as the basis for this book. I am grateful for the many enthusiastic responses it generated, and in particular for the feedback I received from Ronald Meester and Rein Nobel of the VU Amsterdam university. The book before you is an extensive reworking of my Dutch book, made possible with financial support from the Dutch LIME Fund. Many thanks to Peter Shenks, Kate Elliott and Arnoud Vrolijk for their valuable suggestions and their relentless zeal in turning my "Dutchisms" into readable English.

Steven Tijms
Leiden, 2020

Contents

Part One

The Birth

of

Probability

1 The Silence of Antiquity

The Modernity of Probability

Modern life is pervaded by the concept of probability. The weather forecast on television tells you that there is a 50% probability of rainfall for the next day. The doctor assures you that a test will establish with a probability as high as 98% whether you are carrying the disease. A sunblock manufacturer warns you that it is 300 times more probable that you will die from sun exposure than from a shark attack. And your evil math teacher points out that the probability of you becoming a movie star is a mere one in a million.

While to us the concept of probability seems to have always existed, not so long ago things were very different. Neither the ancient Greeks nor the Romans had a word for the probability of an event, and it was not until the second half of the seventeenth century that the mathematical concept of probability emerged.

It is quite astonishing that throughout antiquity apparently no one ever thought of calculating the probability of an event. It is true that the Romans did not make any significant contributions to the field of mathematics, but the Greeks certainly did. They raised mathematics to Olympian heights. The names of Pythagoras, Euclid, and Archimedes are still known all over the world, and the names of many more live on in mathematics up to the present day.

Several explanations have been put forward: that the conservative and agrarian Greco-Roman society would have lacked an economic incentive to calculate probabilities; that the ancient mathematicians with their Platonic outlook would have focused on the eternal truths of the heavens, ignoring the changeable world that is dominated by chance; or, as is often claimed, that the Greeks and the Romans would have believed that the uncertain future was the domain of the gods and therefore inaccessible or even taboo to ordinary mortals. The only recourse one had was to consult a soothsayer or some oracle.

These explanations are all rather speculative, and each of them should be prefaced with a big *perhaps*. Moreover, they suggest that there was something of a common mentality in Greco-Roman society, as if ancient society was not heterogeneous or subject to change. Can we do better? Let's take a closer look.

The Goddess of Chance

Imagine you're walking the streets of ancient Athens or Imperial Rome, asking passers-by what chance is. They may well answer that chance is a goddess. The Greeks called her *Tychè*, the Romans *Fortuna*. In his famous *Natural History*, the first encyclopedia ever written, Pliny the Elder (24–79) describes her: "*Fortuna* is the only one in the whole world that in all places and at every hour and from everyone's mouth is invoked and called by her name. [...] Most people consider her to be fickle and completely blind, with no fixed abode, unstable, uncertain, and changeable, and a benefactor of those who do not deserve it. Every loss and gain is attributed to her and she alone keeps both pages of the books of mortals. So much of our life is subject to chance that we hold chance to be a goddess [...]."

A fickle and blind goddess is what many believed her to be, and a mighty goddess. Fortuna was the only one of the pagan gods to survive both the decline of the Roman Empire and the rise of Christianity and to live on in the popular belief of the Middle Ages.

A Down-to-Earth View

Not everyone believed chance was a divine power. Pliny the Elder himself clearly believed that the mortals themselves had made chance into a goddess. In the same vein the philosopher Epicure dismissed the blind and fickle goddess as a superstition of the masses. And although the great philosopher Plato considered chance to be something supernatural, his famous pupil Aristotle explicitly denied that there was anything divine about chance.

So a substantial number of the ancients, especially the more enlightened ones, simply understood chance as everything that just happened, without any reason or cause. Some believed that chance existed; others, like Democritus, did not. One philosophical school in particular had a surprisingly modern view of chance: The New Academy.

The New Academy

The Academy was founded by Plato in 387 BCE just outside the walls of Athens. The Academy's curriculum included not only Plato's philosophy, but also disciplines such as mathematics and linguistics. Aristotle studied there for almost twenty years before founding his own school, the Lyceum.

In the third century the philosophy taught at the Academy took a radical turn. The philosophers in charge of the school were no longer adherents of Plato's theory of the eternal and unchanging Forms, but skeptics who denied that one could ever be sure of anything. They based themselves on the famous dictum of Socrates "not to think that you know which you actually don't". To distinguish these skeptic philosophers from the old-school Academics their philosophy was referred to as "the New Academy".

They fiercely opposed the dogmatic Stoics, nowadays still known as the imperturbable proponents of the doctrine that everything is predestined by Fate. One philosopher in particular made his living criticizing the Stoics, the charismatic and in his time famous Carneades.

Carneades

At a young age, Carneades (214–129 BCE) had left his hometown of Cyrene, a Greek city in present-day Libya, to study philosophy in Athens. Above all, he meticulously read the writings of the important Stoic philosopher Chrysippus (279–206 BCE). Chrysippus had been training to become a long-distance runner until one day he decided to give up athletics and become a philosopher. Every day he wrote about five hundred lines of text. Upon his death, his oeuvre contained no fewer than 705 books.

Obviously, Carneades must have been an ardent reader. It is told that the philosopher could be so deeply immersed in a book that he had to be fed by his housekeeper and concubine Melissa. But instead of becoming a loyal supporter of Chrysippus, he became his most formidable opponent.

Carneades was a sharp-witted thinker and a very charismatic speaker. He spent his entire life fighting Chrysippus's Stoic philosophy. Parodying the words that Stoicism would not have existed if it hadn't been for Chrysippus, he joked about himself: "Where would I have been without Chrysippus?"

Nothing has survived from Chrysippus' enormous oeuvre, except for a few lines that are quoted by later writers and a few fragments from worn out papyrus rolls. Gone also is his fame. And for his part, Carneades, in the footsteps of his idol Socrates, never wrote anything down. Luckily his disciples did.

Clitomachus, Carneades' most loyal disciple, wrote many books in which he expounded upon his teacher's views. These books are now lost too, but they were studied by the Roman lawyer and statesman Marcus Tullius Cicero (106–43), who was a supporter of the New Academy. Thanks to Cicero we know about Carneades' surprisingly modern view on chance.

Shortly before Cicero was murdered by soldiers of Mark Antony, he wrote a short book about divination, the art of foretelling the future. In the book his brother Quintus defends the Stoic view, while Cicero himself expresses the criticism from the New Academy. Not an original thinker himself, Cicero makes use of Carneades' arguments known to him from his reading of Clitomachus.

Fate or Chance

Divination was widespread in the Greco-Roman world. From priests inspecting the intestines of animals to interpreters of predictive dreams, there were soothsayers who could interpret signs and predict the future as it related to just about everything. In ancient Rome official priests called Augurs interpreted the will of the gods by studying the flight of birds. The Augurs were consulted before a major undertaking was launched, such as matters of war and commerce.

Many Stoic philosophers considered divination to be a special gift or a true art. Divination revealed to the divine soul what Fate had in store for us. Chrysippus is known to have written two books on divination and one on predictive dreams. Carneades, on the other hand, strongly opposed the notion of divination. Whatever the Stoics ascribed to Fate according to Carneades happened by chance.

Unfortunately, Carneades and Chrysippus never met. Carneades was eight when Chrysippus died. But imagine that they had and that we are now in ancient Athens. The day is coming to an end, though the summer heat still hangs over the city like a blanket. In the cool shelter of the *Stoa Poikilè*, the famous colonnade from which the Stoics took their name, the two philosophers meet.

CARNEADES *Dear Chrysippus, what does Fate have in store for us today? Or should I consult a soothsayer or clairvoyant instead of you?*

CHRYSIPPUS *Do I detect a trace of sarcasm? Do as you please. There's nothing happening that is not in the hands of Fate.*

CARNEADES *Superstition and old wives' tales!*

CHRYSIPPUS *Call it whatever you want. You know very well that we Stoics aren't easily provoked. The course of events in the Universe teaches us that Fate is the cause of everything. Present, past, and future.*

CARNEADES *Dear Chrysippus, tell me, what is the point of talking about Fate, when everything can be explained by chance and Nature, without any recourse to Fate?*[a]

CHRYSIPPUS *If that were true, I would like to know how you explain that so many predictive dreams come true! How can this be explained otherwise than that these dreams show us our destiny?*

CARNEADES *You're talking about your new book about dreams? That book is itself a confused dream!*

CHRYSIPPUS *And yet, dear Carneades, all those dreams in my book came true...*

CARNEADES *Dear Chrysippus, we dream all night long and almost no night passes that we don't sleep. Is it then surprising that sometimes something we have dreamed actually happens?*[b]

CHRYSIPPUS *But so many dreams come true, it can't be a coincidence* ...

CARNEADES *Tell me, is there anything as uncertain as rolling dice? Yet there is no one who, as long as he keeps rolling the dice often enough, will not throw a Venus* [This was the highest value one could throw] *once, yes, sometimes even two or three times in a row. Would we, like a bunch of superstitious fools, say that this is because Venus gives the dice a boost rather than that it is a matter of pure chance? And the same is true with predictive dreams!*[c]

CHRYSIPPUS *And yet all those predictions of soothsayers and clairvoyants would not come true so often if they did not understand the art or possess the gift to foresee our destiny.*

---

[a]*De fato*, III 6.

[b]*De divinatione* II 59, 121.

[c]*De divinatione* II 59, 121.

CARNEADES *Dear Chrysippus, don't forget that Socrates is my hero! So tell me, what is a prediction other than knowledge of a future event based on expertise? Like a doctor's prediction of the course of his patient's illness?*[d]

CHRYSIPPUS *That sounds plausible.*

CARNEADES *Thank you. But there are also future events for which there is no special expertise. Like finding a hidden treasure or receiving a large inheritance. And these are the kind of events that soothsayers predict, aren't they?*[e]

CHRYSIPPUS *I can't deny that.*

CARNEADES *But if there's no special expertise, then there's no special reason for these events.*[f]

CHRYSIPPUS *No, but...*

CARNEADES *... but an event for which there is no special reason is called a chance event! And how can anyone predict a chance event? What is divination other than making a guess? As the playwright Euripides once put it: the best soothsayer is the one who makes a good guess.*[g]

CHRYSIPPUS *And yet most predictions come true...*

CARNEADES *Maybe so, maybe not, as we skeptics tend to say. But chance events cannot be predicted. After all, how can there be foreknowledge of events for which there is no reason?*[h] *A chance event is nothing but an event that happens in such a way that it could also happen in a different way or not at all.*[i] *Even a god, my dear Chrysippus,*

---

[d]*De divinatione* II 5, 14.

[e]*De divinatione* II 5, 14.

[f]*De divinatione* II 7, 18.

[g]*De divinatione* II 5, 12; quoted by Plutarchus, *De orac. defect.* 432C.

[h]*De divinatione* II 6, 15.

[i]*De divinatione* II 6, 15.

cannot know in advance future events that happen by chance. So unpredictable they are ...[j]

CHRYSIPPUS *And yet...*

CARNEADES *Yes, I know, Chrysippus. Whether it is chance or Fate that rules, we'll never agree. After all, we're philosophers. Come, the evening is still young. Let's go to a feast.*

The above dialogue never took place, but all the arguments against divination and predictive dreams in particular are taken from Cicero's book. More than two thousand years later these arguments have lost none of their modern bite.

Cicero, or rather Carneades, argues that chance events, like finding a hidden treasure, are not predictable because they have no special reason. He quotes a popular verse from a lost tragedy of Euripides: "the best soothsayer is the one who makes a good guess". Even the gods do not know future coincidences in advance. Clearly, the New Academics did not subscribe to the popular belief that the uncertain future was the domain of the gods.

Predictive Dreams

In ancient times, the belief that dreams could predict the future was widespread. In the *Odyssey* (book XIX 559–67), for instance, Penelope tells Odysseus that there are two kinds of dreams: illusory dreams that come through gates made of ivory and truthful dreams that come through gates made of polished horn.

But not everyone shared this popular belief. In his short treatise *On Divination in Sleep*, Aristotle maintains that if such dreams were really predictive they should have been sent by the Gods. But since only mediocre people received these dreams and not the best and wisest of men, most of these predictive dreams should be explained as coincidences, like when someone goes for a walk and during his or her walk a solar eclipse

[j] *De divinatione* II 7, 18.

occurs. The two events just happen to coincide, without any hidden cause or deeper reason.

Still, predictive dreams can be pretty bizarre. Say that you have a vivid dream about a building being on fire. The next day, only a few blocks away, a house is ablaze. Presumably you are not a pyromaniac and this is not a self-fulfilling prophesy. Nevertheless, you may very well have the feeling that your dream was some kind of "precognition". Many people who have had a predictive dream feel this way. It is simply "too amazing to be a coincidence".

The Stoics thought so too. They believed that dreams could predict the future. The future was predestined by Fate and in dreams the divine soul had access to what was about to happen.

Carneades thought differently. These seemingly predictive dreams, he explains, are not predictive at all. Once in a while a dream happens to come true, but that is because we have dreams all night long and almost every night we sleep. The truth is that it would be exceptional were we not to have, every once in a while, a dream that coincides with reality solely by chance.

The Law of Truly Large Numbers

Carneades' explanation of predictive dreams is an early example of an important law of probability. In the 1980s, mathematicians Frederick Mosteller and Persi Diaconis baptized this law as *the law of truly large numbers.* It states that in a large enough sample or population highly improbable events will happen, just by chance.

Accordingly, life is full of more or less "miraculous" coincidences, as Carneades observed two thousand years ago. In Cicero's words: "Chance is responsible for more miracles in all ages and all over the world than only the ones she has performed in dreams".[k]

To make his point, Carneades draws an analogy between having dreams and throwing dice. Having a predictive dream corresponds to throwing a *Venus*, the highest outcome in ancient dice games. There is no one who does not throw a Venus at least once, and sometimes even two

[k]*De divinatione* II 71, 147.

or three times, if he or she often rolls the dice. This does not happen because the goddess Venus gives the dice a boost, as fools believe, but through chance alone.

It is easy to see why this is so. If you roll the dice only once or twice, not throwing a Venus is a pretty safe bet. But if you roll the dice many times, it will be hard not to throw a Venus at least once. And it is the same with predictive dreams. These too happen simply by chance. So there's no need to believe in Fate or divine inspiration. The law of truly large numbers, as we will show in more detail in Part Two, explains it all.

An Aleatory Model of Chance

If not a goddess, what then is chance according to the New Academics? Chance is, quite simply, short for a chance event. And a chance event is, in the words of Cicero, "an event that occurs in such a way that it at the same time could have occurred in a different way or not at all".[l]

This definition strongly evokes the image of a die with its different outcomes. It was no coincidence that Carneades explained predictive dreams through throwing dice. In the words of Cicero, "nothing is so uncertain as a cast of dice".[m] For Carneades, rolling dice must have been the epitome of chance. Nowadays we would say that a die is a *random number generator*, a device that generates a sequence of outcomes that cannot be reasonably predicted better than by chance.

It is no coincidence that probability theory, which has its applications today in many fields of science and many areas of life, had its origin in gambling with dice. Rolling dice is an easy way of generating random outcomes. And the games of chance that captivated the hearts and minds of the ancient Greeks and Romans were … dice games.

---

[l]*De divination* II 6, 15.

[m]*De divinatione* II 59, 121.

The Popularity of Dice Games

Dice were already in use in the most ancient civilizations. The oldest way of playing dice was with *astragali*. An *astragalus* — or, since it is a Greek word, an *astragalos* — was the knuckle bone from the ankle of a sheep or a goat. The ends of the bone were rounded, but the bone itself had a quadrilateral shape consisting of two wider and two narrower sides. The earliest astragali date from 6000 BCE.

Dice as we know them today also existed in ancient times. The oldest dice date from 3000 BCE and were dug up in Iraq, the former Mesopotamia, and Iran, the former Persian kingdom. Ancient dice have been found from India to Egypt.

In the Indian epic *Mahabharata*, playing dice is the favorite game of kings and nobility. Among the Indian hymns that were collected in the *Rig Veda* (1500–1200 BCE) is the famous *Gambler's Lament*. The hymn is the lamentation of a gambler who has lost his wife's love by squandering all his possessions under the spell of the dice: "Dice are like magic coals. They set hearts on fire while they themselves stay cold". Clearly, the intoxication of gambling is of all times.

The oldest die found in Greece dates from the seventh century BCE. Dice and *astragali* were used side by side by the Greeks and later by the Romans. During Greek feasts, games were usually played with four astragali. The four sides of an astragalus had the values 1, 3, 4, and 6. The opposite sides as a rule added up to seven. The highest outcome, the *Venus*, was when all four astragali had a different value.

Dice games were most often played with three dice. In the case of dice, a Venus was when all three dice were sixes. A special dice board was used on which the dice were thrown. A dice cup was used to shake and throw the dice. Unfortunately, we do not know the rules of the various dice games that were played.

The Romans, if possible even more so than the Greeks, were obsessed with dice games. In Rome alone more than a hundred dice boards have been found. Officially, gambling with dice was only permitted during the feast of the *Saturnalia*, a kind of Roman carnival that was celebrated at the end of December. Only elderly men enjoying their old age were exempt from this restriction. But in practice, no one cared.

The provincial town of Pompeii, which was buried under volcanic ash by the great eruption of Mount Vesuvius in 79 and therefore remained largely intact, provides a good impression of the daily life of the Romans. The wine bar *taberna Lusoria* is preserved, a pub where guests could play dice games while enjoying wine and snacks and afterwards go upstairs to play love games in the brothel.

It goes without saying that no respectable Roman would ever think of visiting such pubs, that is, with the exception of the unscrupulous emperor Nero, who enjoyed hopping incognito from pub to pub in the middle of the night. But that doesn't mean that well-to-do Romans didn't play dice. In higher circles gambling with dice was a popular pastime as well.

The historian Suetonius (69–140) tells in his *Lives of the Caesars* how Emperor Augustus played dice all year round. He quotes from a letter of the emperor to his later successor Tiberius: "During the meal we played dice like old men, not only today but also yesterday. Every time someone threw the Dog [*a one*] or a six, he had to put a denarius in the middle for each astragalus. The pot was for the one who threw a Venus".

The learned emperor Claudius even wrote a book about dice games. Unfortunately, it was lost. Claudius had the reputation of being a passionate gambler. In his carriage, Suetonius writes, a dice board was installed in such an ingenious way that he could play dice during his travels without being bothered by bumps in the road.

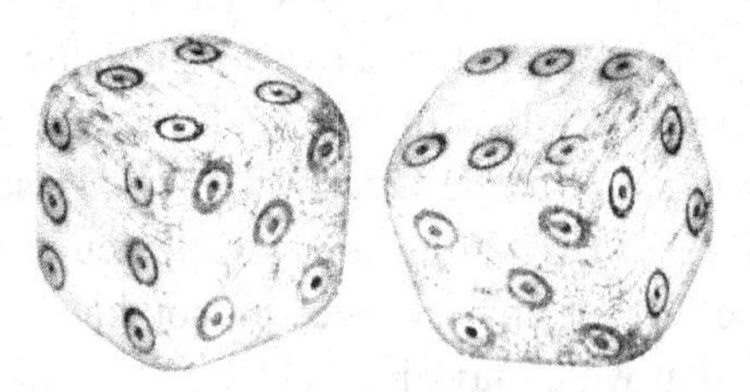

In the Mind of an Ancient Gambler

Contrary to what is often posited, all the ingredients for the development of some elementary probability calculations were actually there. A substantial part of the Greco-Roman intelligentsia did not think of chance as the inaccessible domain of a fickle goddess or some kind of divine power at all. In particular, the philosophers of the New Academy held a pretty modern notion of chance. Their concept of chance was based on the model of rolling dice. And gambling with dice was, as it happens, an immensely popular pastime in ancient times.

So in addition to having a receptive mindset there was also a clear incentive for them to calculate probabilities. After all, some basic knowledge about the probability of the outcomes of rolling dice would give gamblers a highly desirable edge over their opponents.

Imagine you are an avid gambler living in the age of the Roman emperors, visiting a gambling house somewhere in ancient Rome. You sit down with your opponents at a table in the back, away from the light. The dice cup and the dice board are in front of you. In the middle of the table is a large sum of money. The game can now begin.

All of those at the table are experienced gamblers. Would you and your opponents really think of the outcomes of the game as the whims of Fortune? Would you set all your hopes on the favor of a fickle goddess?

In the short run, the ways of chance are erratic and unpredictable. Everything is possible. Who doesn't pray for some luck? Yet your experience sets you apart from the superstitious amateur. You do not believe in a fickle goddess that rules over the outcomes of the dice, changing her mind all the time.

The die itself generates its outcomes, without any preference for one of its six possible outcomes. Clearly, gamblers like you very well understood the notion of a fair die. If not, why would there be so many loaded dice in circulation, as modern archaeological finds have revealed? With every roll of a fair die, every single one of its six possible outcomes had equal rights, so to speak.

But what about the outcomes of three dice? When rolling three dice the possible outcomes, ranging from a sum of 3 to a sum of 18, no longer have equal rights. You know from experience that with three dice some outcomes are easier and others are harder to throw. And the harder they are to throw, the less often they will tend to occur.

Through long experience you have developed a kind of feeling for what we nowadays call probability. There is a logic behind dice games. If only you knew *why* exactly some outcomes are easier or harder to throw than others or, even better, *how much*. It has to be something mathematical, but every time you try to figure it out, all those different numbers and combinations make your head spin!

The Fragments of Antiquity

We have let our imagination run free. But why shouldn't anyone in Greco-Roman times have had these very same thoughts? Gambling with dice was hugely popular. Isn't it unlikely that, in the ten centuries that spanned the flourishing of Athens to the decline of the Roman Empire, no one ever wondered why some of the outcomes of rolling three dice were easier and some more difficult to throw? That would have been almost as unlikely as no one ever having had a predictive dream!

Of course, it is quite possible to play a game of chance for entertainment or even for profit with only some vague notion of the relations between the different outcomes. There are lots of things people do without ever thinking about the rationale that underlies them. Yet there also have always been people who start asking themselves questions. And if one people made a sport out of questioning everything, the ancient Greeks were that people. To this attitude we owe mathematics and Western philosophy.

So why would there never have been a mathematician or a mathematically gifted gambler who had found a solution to the above questions? The silence of the ancients is not an argument for silence.

Quite likely some gamblers knew how to calculate the odds. But even if they had written down their findings — the art of printing was still unknown and memorizing still played a large role in transmitting knowledge — and even if these gamblers had been as famous as Carneades once was or Archimedes still is, it is quite doubtful that these writings would have survived to the present day.

By far the greatest part of what was written in Classical antiquity was destroyed either by barbarians or by Christians, if it hadn't already been erased by the ravages of time or fire. Only a few fragments have luckily been preserved from the writings of many of the greatest poets, scientists, and philosophers of Greco-Roman antiquity. Our knowledge of antiquity is literally fragmentary and highly incidental.

As popular as dice games were, we know next to nothing about the rules of the various games that the ancient Greeks and Romans played. Not even a fragment has escaped oblivion. Wouldn't it have been even more of a lucky coincidence if a text with calculations of the odds had survived the ravages of time?

2 Dice and Odds

A Medieval Ovid

Barren Times

Compared to the Olympian heights that the ancient Greeks had reached, the mathematical landscape in the medieval West was relatively flat and barren. At a basic level, mathematics was taught as part of the seven liberal arts, a kind of preparatory education program that granted access to the major studies at the university.

The mathematical part of the program consisted of some basic knowledge of arithmetic and geometry, some mathematical knowledge of music, especially the scales of church music, and some knowledge of astronomy, especially the circular movements of the celestial spheres and the mathematical principles of astrology.

If you think you could start studying mathematics at an academic level after graduating from this preparatory program, then you are mistaken. Medieval universities were very different from the modern strongholds of science and scholarship. At a medieval university one could only study theology, law, or medicine. So if you wanted to become a professional mathematician instead of a theologian, a lawyer, or a doctor, you were born in the wrong era. An academic study of mathematics, as we know it today, did not yet exist.

The Popularity of Dice Games

The fervor of the medieval gambler, on the other hand, was by no means inferior to the enthusiasm of his ancient predecessor. Like in antiquity, gambling with dice was a popular pastime. There even existed guilds that were specialized in the manufacture of dice. A very popular game was Zara, an early forerunner of Craps, a modern game of chance invented in the nineteenth century that is still very popular in the casinos of Canada and the United States.

The chronicler Odericus Vitalis (1075–1143) writes in his *History of the Church* that even clergymen and bishops could not resist the temptation of dice. Over a century earlier, Bishop Wibold of Cambrai (around 965) had devised a dice game especially for the clergy. Each combination of three dice represented a certain virtue. The clergyman who collected all the virtues first won. The bishop, of course, hoped to divert the attention of the gambling clergymen from worldly affairs. Whether he had any success is a good question. The expected number of rolls that is needed to win this game is more than 1650.

Gambling with dice was usually practiced by rolling three dice and placing bets on the possible outcomes. As mentioned above, some sums are more difficult to roll than others when using three dice. For instance, it is easier to roll a sum of 10 than a sum of 13. So it would be helpful for gamblers if they knew which sums tend to occur more frequently and which tend to occur less frequently. And how much more or less frequently. As in Antiquity, it is hard to imagine that in the Middle Ages gamblers never asked themselves this question. Around the middle of the thirteenth century, for the first time that we know of, an exact calculation of the odds was made for each sum of three dice. Not in a mathematical treatise but in a lengthy poem about love.

Ouidius naso poeta

The Old Woman

In several manuscripts dating from the thirteenth and fourteenth centuries a long Latin poem entitled *The Old Woman* (*De vetula*) has been handed down to us. Its presumed author was the famous Roman love poet Ovid (43 BCE–17 CE), whose poetry was widely read in the Middle Ages.

According to the medieval preface, the poem had been recently discovered in Tomis, an ancient town on the coast of the Black Sea. Ovid had spent the last years of his life in exile in Tomis, banished from Rome by decree of Emperor Augustus. When his tomb was opened, the poem was found in an ivory receptacle, which had protected the manuscript from the ravages of time.

The poem describes how Ovid experienced a kind of moral metamorphosis during his life. In his early years the poet falls in love with a beautiful young woman. One day they agree to meet each other when it is dark and so this comes to pass — at least, that is what the young poet thinks. In reality the lady is not his adored mistress, but an ugly old woman. In the end the young woman marries someone else.

Twenty years later, the husband of the woman he once adored dies. Ovid marries her after all, but the young beauty is now an old woman herself. The bitterly disappointed poet bids his amorous life farewell and decides to devote himself entirely to the study of mathematics, music, and philosophy. Ovid becomes a kind of cleric *avant la lettre* and at the end of the poem he even predicts the coming of Christ on earth.

Pseudo-Ovid

Obviously, the poem with its Christian morals is a medieval fabrication. If Ovid had really been the author of the poem, this would revolutionize our picture of the Roman love poet, to say the least. The famous fourteenth-century love poet and scholar Petrarch wrote: "It is a mystery to me who attributed this poem to Ovid and why. Unless someone was perhaps looking for fame by linking this obscure work to a famous name and, as often happens, laid the eggs of a peacock under a chicken, but then the other way around [...]."

In spite of Petrarch's biting comment, the poem enjoyed considerable popularity during the Middle Ages. As many as sixty handwritten copies are preserved. Early references to the poem by other authors imply that it was composed around the middle of the thirteenth century.

A Poet Versed in Mathematics

Who the real author of the poem was we do not know. Fifteenth-century humanists point to the French cleric Richard de Fournival (1201–60). De Fournival once held a major position at the cathedral of Notre Dame in

Amiens. He is also known as the author of love poems which were inspired by the poems of Ovid. So he was both a love poet and a priest.

The catalog of his library shows that he was at home in many different fields of knowledge, including mathematics. That the author of the poem must have been well versed in mathematics is shown by a remarkable passage.

The poet confesses that in his younger years he used to gamble. Apparently, he ceased to do so and repented, for his confession is followed by a long sermon against playing dice for money. Only fools put their possessions at the mercy of chance. But then the poem takes an amazing turn. The poet priest announces that he will reveal some knowledge which he deems quite helpful for players of these games. He systematically calculates the odds of all possible sums of three dice!

The dice game which de Fournival had in mind was presumably Zara. This popular game was played by rolling three dice. Everything revolved around the sum of the pips on the three dice. For example, if you roll three pips, the sum is 3. If you roll a one, a two and a four, the sum is 7. The game was played in all kinds of variations, but the basic principle was simple. Before throwing the dice, each player called out a number. Then the players took turns rolling the dice. The first player to throw his number won.

De Fournival's Principle

In the heat of the game, most of us only pay attention to the whimsical outcomes of the individual rolls of the dice. Sometimes you get lucky, sometimes you don't. The outcomes could just as well be ruled by the fickle goddess Fortune or, if you prefer, Lady Luck.

A single die has no preference for any of the six possible outcomes, but if you roll more than one die, it's a different story. Some sums of the dice clearly dominate others. But why should blind Fortune favor certain outcomes over others? Apparently the unpredictability of the dice has a logic of its own.

It is easy to see that the sum of two or more dice can often be rolled in more than one way. For instance, if you roll two dice, you may throw the sum of 5 by throwing a one and a four, but also by throwing a two and a three. From this observation Richard de Fournival derived a principle that is as simple as it is fundamental. The more different combinations produce a certain sum, the easier it will be to throw this sum and the more often it will tend to occur.

It is not the rule of Fortune, but a mathematical principle that lies at the basis of dice games. And this means you can actually calculate your chances! Let's see how this works.

How to Calculate the Odds

The sum of the pips on three dice can have sixteen different values, ranging from 3 pips (if you roll three ones) to 18 pips (if you roll three sixes). Some of these sums are more likely to occur, others less likely. To know which sums are more likely to occur and which sums are not, you simply count for each sum the number of possible combinations that produce that sum.

For instance, only one combination produces a sum of 3, a combination of three ones. We will notate this combination as {1, 1, 1}. Equally, there is only one combination that produces a sum of 4, a combination of two ones and one two, or {1, 1, 2}. But there are *two* combinations that produce a sum of 5: {1, 1, 3} and {1, 2, 2}. And *three* combinations that produce a sum of 6: {1, 1, 4}, {1, 2, 3} and {2, 2, 2}. And so on.

In total, there are 56 different combinations of three dice. Since these combinations are obviously unevenly distributed among the sixteen possible sums of the dice, some sums are easier to throw and others harder.

But wait! If the number of different combinations tells us how often a certain sum tends to occur, then it is clear that the sums of 3 and 4 should tend to occur equally often. After all, both can only be produced by one combination. And yet 4 tends to occur on average three times as often as 3. Why is that?

We have overlooked something crucial. The combination that produces a sum of 3, the combination {1, 1, 1}, can be thrown in one way only. But this does not hold true for the combination that produces a sum of 4. The combination {1, 1, 2} can be rolled in three ways: 1 1 2 – 1 2 1 – 2 1 1. After all, we are rolling three *different* dice.

If you have trouble envisioning this, just imagine that each of the three dice has its own color. For example, one of them is blue, one yellow, and one red. Now it is easy to see that the combination {1, 1, 2} can be achieved in three different ways, since each of the three dice can take the value of 2.

In mathematics these variations of the same combination are called *permutations*. Permutations are combinations in which the arrangement of the elements matters. We will simply call them *ordered combinations*. We will reserve the unqualified term *combination* for cases where the arrangement does not matter.

It is easy to see why the number of *ordered combinations* must be much higher than the number of 56 possible *combinations* of three dice. Each of the six possible outcomes of the first die can be combined with the six possible outcomes of the second die. And each of the resulting 36 combinations can in turn be combined with any of the six possible outcomes of the third die. So the total number of possible outcomes is not 56, but $6 \times 6 \times 6 = 216$.

Leibniz's Blunder

It may be a comforting thought that in probability theory even great mathematicians sometimes make elementary mistakes like the above. Take, for example, the one made by famous German mathematician and philosopher Gottfried Wilhelm Leibniz (1664–1716).

Leibniz was a great mathematician. Independently from his English archrival Isaac Newton (1643–1727), he developed differential and integral calculus. At the age of fifty, Leibniz wrote in a letter that "with two dice it is as easy to throw twelve pips as it is to throw eleven pips, because both can only be achieved in one way; but it is three times as easy to throw seven, because this can be achieved by throwing 6 and 1, 5 and 2, or 4 and 3".

But the great Leibniz was wrong. He overlooked that he was rolling two *different* dice and that, as a result, it is in fact twice as easy to throw the sum of 11 (5 6 – 6 5) as it is to throw the sum of 12 (6 6). For the same reason, it is not three but six times easier to throw the sum of 7 (1 6 – 6 1 – 2 5 – 5 2 – 3 4 – 4 3) than to throw the sum of 12. So it turns out that even great mathematicians can get it wrong when it comes to calculating the odds!

The Table of Odds

Richard de Fournival was clever enough not to make the same treacherous mistake Leibniz had. The text of the poem may be somewhat cryptic, but the mathematics behind it is clear and de Fournival proceeds in a systematic way.

For each sum of three dice he determines the number of ways it can be thrown — that is, not merely the number of *combinations* but the total number of *ordered combinations*. As a result, he is able to indicate for each sum of the three dice how many of the 216 possible outcomes produce it. Put in a table, it looks like this:

Tabula III.

Qvot Punctaturas, et qvot Cadentias habeat qvilibet numerorū compositorum.

3	18	Punctatura	1	Cadentia	1
4	17	Punctatura	1	Cadentiæ	3
5	16	Punctaturæ	2	Cadentiæ	6
6	15	Punctaturæ	3	Cadentiæ	10
7	14	Punctaturæ	4	Cadentiæ	15
8	13	Punctaturæ	5	Cadentiæ	21
9	12	Punctaturæ	6	Cadentiæ	25
10	11	Punctaturæ	6	Cadentiæ	27

The first two columns present the different sums that the three dice can produce. Between the two columns there is an inverse symmetry. The sum of 3 is paired with the sum of 18, the sum of 4 with the sum of 17, and so on. In the next chapter, we will return to this remarkable symmetry.

The column in the middle indicates for each sum the number of possible *combinations* (*punctaturae* in Latin). If you add up all of the numbers in the column and double the result to account for the two sums represented, you will notice that there are 56 possible combinations of pips.

The last column indicates the total number of possible *ordered combinations* for each pair of sums, that is, the number of possible outcomes when the arrangement of the dice is taken into account. If, like above, you add these together and double the result, you will see that there are 216 possible *ordered combinations*.

It is interesting to note that the poet calls these outcomes *cadentiae*, which is Latin for the falling of the dice. The word *chance* derives from this Latin term. So *chance* originally meant the "fall", i.e. the "outcome of rolling the dice". In modern probability theory these "fallings" are referred to as *possible outcomes*, also when random processes other than rolling dice are concerned.

Obviously, this table is of great help to gamblers who play dice for money. For each sum of the three dice, a gambler can infer from the table how many possible ways it can be thrown. In the table, the highest number of possible outcomes, 27, corresponds to the sums of 10 and 11. The lowest number, with only one possible outcome, corresponds to the sums of 3 and 18.

Earlier we said that if you roll three dice often enough you'll probably notice that sum 10 is more common than sum 13. Now you can see why. There are 27 possible outcomes that produce the sum of 10 and only 21 possible outcomes that produce the sum of 13. So it is easier to throw the sum of 10 than the sum of 13, and therefore the sum of 10 tends to occur more often than the sum of 13. As the poet says, some numbers are "stronger" than others.

Of course, these calculations are only valid if all of the 216 possible outcomes are equally probable. As long as we do not use loaded dice, they are. As we said before, a die has no preference for one of its six possible outcomes and therefore all 216 possible ordered combinations of three dice share the same probability of 1 in 216.

Calculating the Odds

Listing and counting all possible outcomes is the first step towards calculating probabilities. Yet Richard de Fournival never mentions the word *probability* in its modern sense. Instead, he speaks of "stronger" or "weaker" throws. Or throws that occur "more often" or "less often". Although a term for mathematical probability did not yet exist, one could say that an unarticulated concept of probability was already present.

Anyway, gamblers didn't calculate probabilities. They calculated what we nowadays call *the odds*. Suppose you bet on 10 and your opponent on 13. Then de Fournival's table tells you that your chances of winning are 27 to 21. In other words, the odds are 27 to 21 in your favor.

An Ocean of Silence

In the history of probability theory, Richard de Fournival's calculations are like an island in an ocean of silence. It is more than three centuries before we meet someone else who correctly calculates the odds of dice games. But it would be very remarkable if de Fournival was the first and only one in three centuries to have precisely calculated the odds. Presumably the knowledge of calculating the odds had been, and still was, passed on through private contacts between gamblers (as probably was also the case in ancient times).

This also explains why some of these gamblers had difficulty performing the calculations correctly. Luckily the gamblers you'll meet in the next chapter had the best teacher they could wish for: the famous Italian mathematician, astronomer, and physicist Galileo Galilei.

The Divine Mathematician

The Book of Nature

With the birth of Galileo Galilei (1564–1642) a new era dawned. In the fifteenth century, European mathematics had begun to develop rapidly, partly because of the invention of book printing and partly because Arabic numerals (0, 1, 2, …) gradually replaced the Roman numerals (I, V, X, L, C, D and M). At the time of Galileo's birth, the modern numeral system was widely in use. Yet science as we know it today was still practically non-existent. Nature was the domain of philosophy, at least as long as the philosophers did not contradict the truths of faith. The ancient Greek philosopher Aristotle was regarded as an authority. Latin translations of his books were read and studied at every university in Europe.

Galileo radically broke with this tradition. "Philosophy," he wrote, "is written in the large book that lays wide open before our very eyes". Galileo meant the universe itself. And the book of nature is written not in Latin, but in the language of mathematics "without whose help it is impossible to comprehend in human terms a single word of it".

Galileo's words are widely regarded as a manifesto of modern science. By the time of Galileo's death, mathematics had become the universal language of the new science, and observation and experiments had replaced tradition and authority in the study of nature. Albert Einstein, who greatly admired his Italian forerunner, called him *the father of modern science.*

The Copernican Model

In 1616 Copernicus' thesis that the earth does not lie at the center of the universe and that we revolve around the sun like any other planet was condemned as "foolish and absurd in philosophy, and formally heretical since it explicitly contradicts in many places the sense of Holy Scripture". Yet Galileo's observations with his telescope strongly favored Copernicus's heliocentric model.

Galileo's observations made him world famous in no time. He observed many things in the Universe that until then had been invisible to the eye. He discovered that the moon is a pocked spherical lump of stone and not the perfect crystal ball many scholars imagined it to be, that four moons revolve around Jupiter, and that Venus, like our moon, has different phases.

And Yet It Moves!

In 1632 Galileo publishes his *Dialogue Concerning the Two Chief World Systems*. In the book, written in Italian and meant for a large audience, Galileo contrasts the Copernican heliocentric view with the geocentric view of Aristotle and the Catholic Church. The book becomes a bestseller, but Pope Urban VIII is not amused, as it is easy to read between the lines which of the two views Galileo favors. What's more, Galileo has put several statements made by the Pope into the mouth of Simplicio, the character that maintains the view that the earth is at the center of the Universe. The *Dialogue* is placed on the Index of Prohibited Books and at the age of 70 Galileo is forced to officially renounce the heliocentric view, or else he will end his life in a dungeon.

On 22 June 1633, Galileo professes: "I, Galileo, son of the late Vincenzo Galilei of Florence, seventy years of age, arraigned personally for judgment, kneeling before you Most Eminent and Most Reverend Cardinals Inquisitors-General against heretical depravity in all of Christendom, having before my eyes and touching with my hands the Holy Gospels [...] with a sincere heart and unfeigned faith I abjure, curse, and detest the above-mentioned errors and heresies".

The deeply saddened Galileo is forbidden to publish books any longer and is placed under house arrest for the rest of his life. After his verdict, he allegedly mutters the famous words: “And yet it moves!”

Five years after the verdict Galileo, who has seen deeper into the Universe than any mortal being before, eventually loses all of his eyesight. In November 1641 he becomes severely ill and suffers from a high fever. Early in 1642, he dies at the age of 77.

The Archduke of Tuscany wishes to erect a marble mausoleum for the famous scientist, but in accordance with papal orders his body is buried anonymously in a small room of the Basilica of Santa Croce in Florence. The present monument in the main part of the church dates only from 1737. It was not until 1992 that Pope John Paul II rehabilitated Galileo Galilei. He openly apologized for the wrongs the Church had done him.

The Divine Mathematician

From a young age Galileo must have been convinced that mathematics is the language of nature. Already as a child he is fond of solving problems mathematically. At age sixteen Galileo is sent by his father, Vincenzo Galilei, who today is still known as an important music theorist, to the University of Pisa to study medicine. Vincenzo wants to keep his son away from mathematics. As a doctor he would earn a lot more money than as a mathematician.

Galileo never completes his medical degree, and after four years he leaves the university to study mathematics, first with Ostio Ricci, the mathematician at the court of the Archduke of Tuscany, and later with Christopher Clavius in Rome. He passionately studies the works of Archimedes, his favorite mathematician and idol.

Galileo soon makes his name as a mathematician. At just 25 he is appointed chair of mathematics at the University of Pisa. Three years later, after his father has died, he moves to the University of Padua where he teaches mathematics, including mechanics and astronomy. He lives in the company of Marina Gamba, his maid, and although they have three children, two daughters and a son, they never marry.

In 1610 the Archduke of Tuscany, Cosimo II de' Medici, invites Galileo to his court. The by now famous Galileo — he is even called *the divine mathematician* — is offered the post of "Mathematician to his Serenest Highness". Galileo accepts the offer. He also is made "first and extraordinary mathematician" at the University of Pisa without having any obligation to teach.

A nice impression of life at the court at that time is given by the editor of Galileo's collected works (1718): "It was customary in that happy period in our city of Florence to hold learned meetings in the homes of gentlemen who did not spend their time on women, horses or excessive gambling, but rather passed their days with scholarly discussions among educated people". Perhaps it was at one of these meetings that the archduke asked Galileo to solve the following problem.

The "Paradox of the Three Dice"

Some diehard gamblers, probably from the circle of the Archduke's friends, had noticed through their long experience that with three dice it is slightly more advantageous to bet on 10 or 11 than on 9 or 12. Yet this ran counter to their mathematical calculations.

According to these calculations, there were just as many combinations that produced 9 as there were that produced 10, 11, and 12. Six combinations for all four sums. How could this paradox be explained? A quick glance at the table of Richard de Fournival will give you the answer, but the challenge is, of course, to find out for yourself where the gamblers went wrong.

Galileo wrote a short essay about the problem under the title *About the Outcomes of the Dice* (*Sopra le Scoperte dei Dadi*). At the start of his essay Galileo states the reason why some sums of the three dice are more advantageous than others. It all comes down to the number of possible outcomes that produce the same sum. The larger the number of these outcomes is, the more often the sum will tend to occur. Exactly the same principle Richard de Fournival formulated some three and a half centuries before. Galileo's next step is to show how to correctly calculate these numbers.

An Inverse Symmetry

Galileo confines his calculations to the sums of 9 and 10. Perhaps you can figure out why. The crux is that the pips on two opposite faces of a die always add up to seven (1 + 6, 2 + 5 and 3 + 4). It follows that the sum of the pips on the opposite faces of the three dice is always 21 (3 × 7).

So if the pips on the uppermost faces of three dice add up to a sum of 10, those on the underlying faces have to add up to a sum of 11 (21 − 10), and vice versa. As a consequence, the number of ordered combinations that add up to 10 and the number of ordered combinations that add up to 11 have to be equal. And the same holds for the sums of 9 and 12 and for all other twin sums that add up to 21. This is why Galileo could confine his calculations to the sums of 9 and 10, leaving aside the sums of 11 and 12. It is also the reason for the inverse symmetry in de Fournival's table.

The Gamblers' Mistake

The gamblers were right that there are as many combinations that add up to 9 as there are combinations that add up to 10. Six combinations — {1, 2, 6}, {1, 3, 5}, {1, 4, 4}, {2, 2, 5}, {2, 3, 4}, and {3, 3, 3} — add up to 9 and six combinations — {1, 3, 6}, {1, 4, 5}, {2, 2, 6}, {2, 3, 5}, {2, 4, 4}, and {3, 3, 4} — add up to 10.

Yet it had escaped the gamblers — but not Galileo — that these combinations differ in the number of ways they can be thrown. Some of them can only be thrown in one way, while some of them can be thrown in more than one way. If you have trouble envisioning this, just imagine that each die has its own color.

The gamblers only counted the number of combinations. Instead, they should have counted the total number of *ordered* combinations. It is a quite laborious task to list and subsequently count all possible ordered combinations of three dice. Galileo therefore uses a clever shortcut.

Galileo's Shortcut

Suppose that all three of the dice thrown show the same number of pips. Then there is only *one* way in which this combination can be thrown. For example, the combination {3, 3, 3} can only be thrown as 3 3 3.

If on the other hand only two of the thrown dice show the same number of pips, then there are *three* possible ways to throw such a combination. For example, the combination {2, 2, 5} can be thrown in the following three ways: 2 2 5 – 2 5 2 – 5 2 2.

Finally, if a combination features a different number of pips on each of the three dice, there are *six* possible ways to throw such a combination. For example, the combination {1, 2, 6} can be thrown in the following six ways: 1 2 6 – 1 6 2 – 2 1 6 – 2 6 1 – 6 1 2 – 6 2 1.

So some combinations can be thrown in only one way, some can be thrown in three ways and some can be thrown in six ways. Now it is not hard to count the total number of ordered combinations that produce respectively the sum of 9 and the sum of 10. Galileo only had to determine to which of the above three "categories" the six combinations of each sum belonged.

Sum 9		Sum 10	
{1, 2, 6}	6	{1, 3, 6}	6
{1, 3, 5}	6	{1, 4, 5}	6
{1, 4, 4}	3	{2, 2, 6}	3
{2 ,2 ,5}	3	{2, 3, 5}	6
{2, 3, 4}	6	{2, 4, 4}	3
{3 , 3, 3}	1	{3, 3, 4}	3
6	25	6	27

The six combinations making up the sum of 9 correspond to 6 + 6 + 3 + 3 + 6 + 1 = 25 possible outcomes (see table). The six combinations making up the sum of 10 correspond to 6 + 6 + 3 + 6 + 3 + 3 = 27 possible outcomes (see table). The odds are therefore 25 to 27 in favor of 10, as we would say nowadays. And this explains why it is indeed slightly more advantageous to bet on 10 rather than 9 and, consequently, on 11 rather than 12.

The Silence of the Middle Ages

Galileo's essay was not published until 1718, when it was incorporated into his collected works. Its influence therefore has been marginal, yet it presents a clear picture of how gambling was the trigger for the first steps towards a calculus of probability.

Galileo proceeded in essentially the same way as Richard de Fournival had, but with far more ease and clarity. Galileo was of course a great mathematician. Yet it is also a reflection of how familiar mathematicians and gamblers — or gambling mathematicians — had become with these kinds of calculations by that time.

Still almost no written testimonies from this period are known to us. A rare instance can be found in a fourteenth-century commentary on a passage from Dante's masterpiece *The Divine Comedy*. The commentator, Jacopo della Lana, explains the logic of the dice game Zara, yet he makes the same mistake as Leibniz and the archduke's gambling friends.

One might wonder why it had been so quiet all these centuries between Richard de Fournival and Galileo Galilei. There are several reasons one can think of. Unlike today, there were no scientific journals and knowledge was often exchanged privately, either orally or in personal correspondence. Printing was only introduced around 1450.

Moreover, contrary to the current publish or perish culture, both mathematicians and gamblers benefited from keeping their knowledge to themselves. For gamblers this is quite understandable, as their knowledge of the odds gave them an edge over ignorant opponents. And the same was true for mathematicians.

In the Middle Ages a lively way of teaching had developed: the public debate. Usually a debate was held between two scholars in front of a mixed and enthusiastic audience. The scholars set each other a number of problems, and the one who succeeded best in solving the opponent's problems was declared winner of the debate. It was therefore not to one's advantage to publish solutions one had found with some difficulty. No doubt more people knew how to calculate the odds correctly, but most likely it remained insider knowledge.

However, everything hadn't been completely silent in the three and a half centuries between de Fournival and Galileo. In the sixteenth century there had been an Italian mathematician who was as nonconformist as Galileo was to be. Going beyond simply calculating the odds, he formulated a general rule to establish a fair proportion between the bets in dice games.

The Gambling Scholar

An Italian Polymath

Girolamo Cardano (1501–76) was one of the most colorful mathematicians in history. Today he is almost forgotten, but during his lifetime he was an international celebrity. The erudite scholar not only read like a madman, he wrote like a madman as well, with 130 books and 111 manuscripts to his name. According to himself, he burned another 177 manuscripts. Apart from treatises on topics from mathematics, he wrote books on the most diverse subjects such as one on the interpretation of dreams and a hymn to Emperor Nero. Yet, of his many books only his autobiography, *About My Life* (*De Propria Vita*), has stood the test of time. The book is a classic in its literary genre. Cardano describes both his many talents and his many shortcomings, down to the most personal details.

The Early Years

Cardano is born in Pavia in 1501. As a baby he survives not only an abortion but also the Plague. His three brothers are less fortunate: they all succumb to the Black Death. His father is a friend of Leonardo da Vinci, who visits him regularly to discuss mathematical problems. For though father Cardano is a lawyer, he is actually more interested in mathematics and astrology. As a result, when Girolamo is barely twelve, his father gives him the first six books of Euclid's *Elements* to study.

At the age of nineteen Girolamo goes to the University of Pavia. As mathematics is not yet an independent study at universities, Cardano begins his academic career by studying medicine and philosophy. He earns a living as a tutor of mathematics. At the university he soon makes a name for himself. He outshines the other students in public debates on the most diverse issues, and when there is something to celebrate he is the life and soul of the party.

Once graduated, Cardano becomes a doctor. Initially, he lives in poverty. Together with his brilliant pupil, Ludovico Ferrari (1522–65), he writes a comprehensive overview, full of the latest insights, of the field of algebra. The book, entitled *Ars Magna* (*The Great Art*) and written in Latin (Latin was still the international language of communication, as English is today), becomes the standard work on algebra and remains so for almost a century. It brings Cardano international acclaim, at least among mathematicians.

HIERONYMI CAR
DANI, PRÆSTANTISSIMI MATHE
MATICI, PHILOSOPHI, AC MEDICI,
ARTIS MAGNÆ,
SIVE DE REGVLIS ALGEBRAICIS,
Lib. unus. Qui & totius operis de Arithmetica, quod
OPVS PERFECTVM
inscripsit, est in ordine Decimus.

HAbes in hoc libro, studiose Lector, Regulas Algebraicas (Itali, de la Cosa uocant) nouis adinuentionibus, ac demonstrationibus ab Authore ita locupletatas, ut pro pauculis antea uulgò tritis, iam septuaginta euaserint. Neq; solum, ubi unus numerus alteri, aut duo uni, uerum etiam, ubi duo duobus, aut tres uni æquales fuerint, nodum explicant. Hunc aũt librum ideo seorsim edere placuit, ut hoc abstrusissimo, & planè inexhausto totius Arithmeticæ thesauro in lucem eruto, & quasi in theatro quodam omnibus ad spectandum exposito, Lectores incitarẽtur, ut reliquos Operis Perfecti libros, qui per Tomos edentur, tanto auidius amplectantur, ac minore fastidio perdiscant.

It also earns him the ire and hatred of the Italian mathematician Niccolò Tartaglia (1499/1500–57), nicknamed *the Stutterer*. In the form of a cryptic poem, Tartaglia has entrusted Cardano with a method for solving third-degree equations. Cardano, however, has published Tartaglia's formula in his book without giving the latter credit. Tartaglia is furious and accuses Cardano of having published the formula under his own name.

The accusation is not entirely true. Tartaglia is not the first to have solved this type of equation. The Italian mathematician Scipione del Ferro (1465–1526) has already done so and, though he has never published the formula, Cardano knows about it.

Nevertheless, a long and vicious correspondence ensues between Tartaglia on the one side and Cardano and young Ferrari on the other. Both sides outdo each other in sarcasm. Eventually, the correspondence culminates in a public debate between Tartaglia and Cardano's pupil, Ferrari. When Ferrari turns out to be Tartaglia's superior by far, the Stutterer prematurely leaves the room.

Until the end of his life Tartaglia, unable to face the humiliating defeat, will maintain that he himself and not "the creature of Cardano", as he consistently calls Ferrari, has won the debate. The Stutterer's resentment will haunt Cardano for decades. It is the first, but not the only, time that Cardano falls victim to the envy of his colleagues.

At the Peak of Fame

Cardano's fame as a scholar and mathematician spreads steadily beyond the borders of Italy. Yet his career reaches truly dazzling heights when, quite unexpectedly, he receives a request to cure the archbishop of St. Andrews. The powerful and important clergyman suffers from severe attacks of asthma.

Cardano travels all the way to Scotland, only to find the physicians of both the French king and the Spanish emperor already present at the archbishop's bedside. But whereas the prominent physicians are mainly engaged in learned discussions among themselves, Cardano extensively observes the patient and prescribes a detailed regime of rest, cleanliness, and regularity. The result is amazing. After many years of asthmatic attacks the archbishop recovers at last.

The news spreads like wildfire throughout the courts of Europe. On his way back home, Cardano is summoned to the court of Edward VI, the fifteen-year old king of England. The king asks Cardano to cast his horoscope. Strange as it may seem, it was quite common in those days to ask a physician for a horoscope, since astrology was still regarded as an auxiliary "science" of medicine. Cardano predicts that the king will rule over his subjects as a wise and righteous monarch. Yet within less than a year the young king dies — a tragic event which, as Cardano later acknowledges, he did not foresee.

There is no end to the royal invitations. Cardano also receives an invitation from the king of Denmark and, what's more, both the mighty French king Henry II, who is married to Catherine de' Medici, and his archrival Emperor Charles V would like to add the now famous physician to their courts. Cardano refuses both requests. He does not want to take sides between the two powerful rivals. Moreover, he treasures his personal freedom and abhors the life of the courtier and all its flattery and intrigue.

Dark Times

Cardano has become a celebrity by now. Once he is back home, wealthy and high-ranking patients visit his practice in large numbers. Still, he is not spared the setbacks of fortune. One of his two sons wanders through Italy playing dice and stealing money, even from his own father. The other son poisons his wife after she confesses that none of their children are his. He is tried and beheaded. Cardano, despite his renown, is weighed down by anguish and grief. Yet things will only get worse.

Cardano once wrote a book about malpractice in medicine. The work caused quite some enmity among many of his colleagues. Haunted by their envy and hatred, Cardano finally settles in Bologna, where he is safe from their conspiracies. However, when Pius V is elected Pope, the political climate changes in Bologna as well. In 1570 Cardano is arrested and imprisoned by the Inquisition.

The reasons have never been fully clarified. It may have been his liberal attitude toward religion, but it will also have displeased the Pope that in the past he published a horoscope of Jesus Christ. Thanks to a few benevolent cardinals, Cardano narrowly escapes the stake. After seventy-seven burdensome days of imprisonment he is released on bail. In exchange for the Pope's grace, he must abstain from publishing and teaching for the rest of his life.

Cardano settles in Rome, near his patron cardinals. Every day the inhabitants of the city witness how the eccentrically dressed old man shuffles with difficulty through the streets and over the bridges of Rome. At home he spends his time at his desk, writing his autobiography. Soon after completing the story of his life, he dies. The superstitious Cardano is alleged to have cast his own horoscope and starved himself for three weeks to die exactly on the day he had calculated.

A Handbook on Gambling

Cardano was not only an erudite scholar, he was also a fervent gambler. When he was short of cash or sought to dispel his grief, he threw himself into gambling. Day in, day out, year after year. Cardano would not be Cardano if he had not written about gambling as well. Between 1525 and

1564 he wrote a booklet in Latin, entitled *About Games of Chance* (*De Ludo Aleae*).

Rather than a mathematical treatise, the booklet is a kind of learned handbook on gambling. It numbers only fifteen pages written in double columns and is divided into short chapters with titles such as “Why I write about Gambling”, “Why Aristotle Condemns Gambling”, “About Fraud in Card Games”, “About Luck in Games of Chance”, and “About the Character of Gamblers”. The chapters are sprinkled with more or less mathematical passages.

For a long time the booklet did not receive the credit it deserves, partly because of Cardano’s fast and capricious way of writing — in the text he let mistakes and later corrections stand side by side — and partly because of the apparent absence of any internal logic. The booklet has often been considered to be no more than a clutter of notes, lacking any cohesion. Only recently has there been growing understanding and appreciation of it.

Criticism of Gambling

Over the centuries, gambling has met with a lot of criticism. The Church denounced the immoral conduct associated with the practice of gambling. The ancient Greek philosopher Aristotle also condemned the typical gambler. In his work about true happiness, the *Nicomachean Ethics*, the philosopher mentions the gambler in the same breath as thieves and robbers. The gambler is greedy and only out for shameless profit. Even his own friends, to whom he should be generous, he squeezes for money.

Cardano devotes a whole chapter to the criticism. He addresses Aristotle’s criticism in particular. The works of the ancient Greek philosopher deeply influenced the traditional philosophy in the Medieval West. In Cardano’s time Aristotle’s philosophy was still being taught at the universities, and the *Nicomachean Ethics* was widely read. As we shall see, Cardano’s handbook on gambling will prove to be considerably more coherent and far less incomprehensible against the backdrop of Aristotle’s philosophy.

Fair Play

In contrast to Aristotle, Cardano is convinced that a morally acceptable form of gambling does exist. For gambling to be fair there must be complete equality between two players. “Equality”, he writes, “is the most important principle of gambling”. In fact, the principle of equality is the common thread in his handbook.

Cardano sketches a vivid picture of common practice: spectators interfering noisily with the game, disagreements resulting in major rows, all kinds of trickery and deceit, and, last but not least, the use of unbalanced dice boards and loaded dice. At the gambling table equality is often hard to find.

For gambling to be fair, the principle of equality should therefore apply to all of the following ingredients: the choice of an opponent, the partiality of the audience, the location, the dice table, and the dice themselves. Yet it is not enough for the circumstances to be equal for both players. Equality should also apply to the dice game itself.

In practice, bets were often to the advantage of the most experienced player and to the detriment of an inexperienced but usually greedy opponent. So how is it possible to achieve equality in games of chance? Surprisingly, Cardano finds the answer in the very same *Nicomachean Ethics*. He does not explicitly mention Aristotle, but in the background the philosopher is unmistakably present.

Aristotle on Equality

It is no coincidence that equality is Cardano’s criterion for a fair game. In book 5 of the *Nicomachean Ethics*, still a classical text in the philosophy of law, Aristotle writes: “Justice between people is based on equality”. Equality means that everyone should get their fair share. It can be achieved in two ways, both of which result in a fair distribution.

First of all, there is a type of equality that is proportional to the merits of people. For example, if you have done twice as much work as someone else, you are entitled to a reward that is twice as big. Let’s say that the total amount of money is 30 silver coins or *talents*, as they were called in ancient times. Then you are entitled to 20 talents, whereas the other should get the 10 remaining talents. One could call this type of fair distribution “the method of proportion”.

There is also a second type of equality that applies when people have equal rights. Take an instance in which two persons each have a right to the same amount of money, but one of them makes an unjustified profit at the expense of the other. In this case, Aristotle writes, one adds up the amount of money they both possess and from this total one takes the middle, that is, half of the total sum.

Let’s say that together they possess 30 talents. Then the middle is 15 talents. Now, to restore the balance, the person who has made a loss gets as many talents as he or she is short of from the middle. For instance, if he or she has 10 talents, he or she gets another 5 talents. The person who has made an unjustified profit loses the same amount. As a result, the balance is restored since both end up with the same amount of money: 15 talents. Cardano himself calls this “the method of the middle”.

The Method of the Middle

Now, Cardano does not say in so many words what constitutes a fair game, but clearly a fair game is one in which neither player has the advantage. In other words, the expected profit and loss should be evenly distributed between the two players. At first glance, it seems only logical to opt for the method of the middle. This is indeed the strategy that Cardano initially chooses. An example may illustrate how Cardano proceeds.

Imagine you are playing a game of chance with only one die. It is of course a fair die. Suppose you bet that you will throw a six and your opponent bets that you will not. Three is the middle of the number of possible outcomes of a die. So if you make a bet to throw at least one six in three throws, and both you and your opponent stake an equal amount of money, the expected win and loss should be distributed equally between you.

This sounds quite convincing, but is Cardano’s method of the middle really correct? The answer is no!

Cardano's Error

To see what is wrong with Cardano's method of the middle we will reformulate his reasoning in terms of probabilities. Of course, the concept of probability was still unknown. For this reason we should not judge Cardano too harshly for his mistake. Our first impulse might very well have been the same: to add up the individual probabilities, in keeping with the number of rolls.

If you roll the die once, the probability of throwing a six is 1 in 6. If you roll the die again, you will have another probability of 1 in 6. And so on. What could be more logical than to just add up these probabilities? The probability that you will throw a six in three rolls then is simply 3 in 6. In other words, the odds are fifty-fifty and this is exactly the equality that Cardano was aiming for. Yet the method is wrong.

It's an elementary mistake to think that one can just add up the individual probabilities. If one does, strange results will follow. Suppose you rolled the die six times. Then, according to the same line of thought as above, your probability of throwing a six would be 6 in 6. In other words, you would be certain to throw a six in six rolls. Yet everyone knows that it is possible to roll a die six times and not throw a six once. What's more, what applies to six also applies to the other possible outcomes of the die.

So even though it seems intuitively a logical thing to do, adding up individual probabilities leads to absurdities. The method of the middle cannot be right. Halfway through his booklet, Cardano, in a moment of insight, realizes he is wrong. The odds of throwing a six in three rolls are not fifty-fifty at all!

How to Calculate the Odds

So how should we calculate the odds? First, we establish the total number of possible outcomes. Next, we count how many of these are favorable and how many are unfavorable. A favorable outcome is a winning one, while an unfavorable one is a losing one. The ratio — Cardano uses the Latin word *proportio* — between favorable and unfavorable outcomes we will call *the odds*.

Let's start with the total number of all possible outcomes. It is simply the number of ways in which the six possible outcomes of each roll of the die can be combined. The 6 possible outcomes of the first roll can be combined with each of the 6 possible outcomes of the second roll. The resulting possible 36 combinations can in turn be combined with each of the 6 possible outcomes of the third roll. So there are $6 \times 6 \times 6 = 216$ different possible outcomes of three rolls with a die. Exactly as many as when rolling three dice simultaneously.

Now we should establish how many of these 216 possible outcomes are favorable. That is, we have to count the number of ordered combinations with at least one six.

The number of possible outcomes with exactly one six is $3 \times 5 \times 5 = 75$. With three rolls there are three ways of throwing exactly one six. Either the first roll, or the second, or the third. In each case, for the other two rolls 25 (5×5) possible outcomes remain.

The number of possible outcomes with exactly two sixes is $3 \times 5 = 15$. Again, there are three ways of throwing exactly two sixes. For the one roll that remains 5 different outcomes are possible.

Lastly, the number of possible outcomes with exactly three sixes is, of course, 1. When added to 75 and 15, you arrive at 91 possible outcomes with at least one six.

An even easier way to calculate the number of favorable outcomes is to first calculate the number of unfavorable outcomes, i.e. the number of all possible outcomes with *no* six at all. For three rolls this number is simply $5 \times 5 \times 5 = 125$. To get the number of possible outcomes with at least one six, one only has to subtract this number from the total number of possible outcomes: $216 - 125 = 91$. In short, the odds aren't fifty-fifty, they are 91 to 125.

The Method of Proportion

To achieve equality when the odds are not equal, the method of the middle will not do. So much is clear. That leaves us with the method of proportion. But how should we apply this method to a game of chance? The learned gambler himself explains.

To achieve equality, the bets of the two players have to be proportional to the ratio between the number of favorable outcomes and the number of unfavorable ones. In other words, the amounts of money the players bet must be in the same proportion as the odds. Then the expected loss and the expected profit for each of the two players will be equal and the game will be a fair game.

For example, imagine you are betting that in two rolls of a die you will throw a six at least once. The total number of possible outcomes for two rolls is 36 (6×6). Some favorable outcomes are, for instance, 1 6, 6 1, 6 6, 6 3, and so on. If you count them all, you will see there are 11 favorable outcomes.

Or, more easily, you calculate the number of unfavorable outcomes first (5×5) and then subtract this number from the total of 36 possible outcomes, which also leaves you with 11 favorable outcomes.

So the odds are 11 to 25 against you. In other words, you expect to win on average 11 of the 36 times and to lose 25 of the 36 times. Cardano's rule states that equality is achieved when the stakes are in exactly the same proportion as the odds. Accordingly, you should contribute 11 talents to the pot, while your opponent should contribute 25 talents.

In this way the expected profit and loss are evenly distributed between the two players, as some simple math shows. Since you win on average 11 out of 36 times, your expected profit in 36 throws is equal to $11 \times 25 = 275$ talents. And since you lose on average 25 times out of 36, the expected profit of your opponent is equal to $25 \times 11 = 275$ talents as well. In short, neither player has the advantage. It is a fair game.

The Honest Gambler and Cardano's Rule

Imagine that there has been a debate in which Cardano opposed an Aristotelian philosopher, the question being whether or not it is possible for one to be an honest gambler. Then, based on the above, Cardano turns out to be the rightful winner. He has defeated the Aristotelian philosopher with the latter's own weapons. If the stakes are proportional to the odds, then according to Aristotle's own definition, the game is fair. So not all gamblers should be put on a par with thieves and robbers.

Cardano presents his rule as a discovery, something no one had thought of before. He writes that many people make unfavorable bets and as a result lose a lot of money. This is because they don't understand Aristotle. Cardano apparently refers to the method of proportion. It is the only time Cardano hints at the ancient Greek philosopher as his source of inspiration. Cardano also writes that he wishes to protect naïve gamblers and therefore does not want to keep his discovery secret (no doubt an allusion to Tartaglia).

The greatest merit of Cardano's rule is that it is a general rule. Once gamblers know the odds, they can make fair bets. Cardano is the first to go beyond individual gambling games. For these reasons, the rule should rightly be named after the colorful gambling scholar.

Laplace's Formula

Cardano's rule is one step away from the classic formula for calculating probabilities. If Cardano had divided the number of favorable outcomes by the total number of possible outcomes, he would have stated the classic formula of mathematical probability. The formula became known in the seventeenth century, but its classical wording comes from the mathematical genius Pierre-Simon Laplace (1749–1827).

Laplace, also known as the French Newton, was one of the greatest mathematicians of all time. When young Napoleon attended the *École Militaire*, Laplace happened to be his examiner. Once Napoleon was emperor of France, he made his former examiner a count of the Empire.

In 1819 Laplace published his *Philosophical Essay on Probabilities* (*Essai philosophique sur les probabilités*), a kind of summary for laymen of his mathematical masterpiece *Théorie analytique des probabilités*, the most influential work on mathematical probability theory until the end of the nineteenth century.

The essay begins with a series of principles of probability. The first principle states that the probability of an event (for instance, throwing at least one six in two rolls) is equal to the number of favorable outcomes divided by the total of all possible outcomes, provided that all possible outcomes have the same probability. In our example, the probability of winning would be 11/36 which is about 30.6%. Likewise, the probability of losing would be 25/36. About 69.4%.

Laplace's definition would become the classical formulation. The expressions *favorable outcome* and *unfavorable outcome* also have become standard terms in the theory of probability. They still reflect the fact that the theory of probability has its origin in gambling with dice.

Why Odds?

Cardano already possessed a special term for all possible outcomes of one or more dice: *circuitus*, literally "going around". Cardano probably meant all the outcomes you can get by rotating the die or dice. Why did Cardano, being so close, not calculate probabilities instead of the odds? The simple answer is that Cardano wrote not a mathematical treatise but a book that was intended for gamblers. Gamblers make bets, and when it comes to making bets odds are far more appropriate. They provide an easy way of calculating the size of your stake.

It is still customary to use odds when people are betting, as for instance on horse races. Children, according to the famous Swiss psychologist Jean Piaget (1896–1980), also find odds easier than probabilities. In other words, at an elementary level odds are not such a bad choice. As we will see, it's the concept of probability, instead, that needs some explaining.

3 The Mathematics of Chance

The Child Prodigy and the Counselor

The Scientific Revolution

In the seventeenth century Europe's intellectual landscape profoundly changed. Historians often refer to the period as *the Age of the Scientific Revolution*. The start of the century had seen Galileo stand trial before the Inquisition for claiming that the earth revolves around the sun. Yet before the century had come to an end, Aristotle's natural philosophy, which had been the church's leading doctrine, was forever relegated to the past. Isaac Newton (1643–1727) was able to explain, on the basis of a few principles, both the natural phenomena on earth and the movements of the celestial bodies. The impact of his theory was enormous.

Mathematics as well experienced a profound transformation. While Cardano had still used laborious abbreviations to denote mathematical operations (for instance, m. stood for minus and p. for plus), modern notation — the one we nowadays learn at school — came more and more into use in the seventeenth century.

Mathematics itself also developed rapidly. The logarithm had only just been invented at the beginning of the century and yet by the 1660s and 1670s Isaac Newton, also one of the greatest mathematicians of his time, and Gottfried Wilhelm Leibniz had both, independently of each other, substantially expanded the mathematical toolkit with differential and integral calculus. This meant a huge step forward. Many difficult mathematical problems could now easily be solved. Moreover, calculus also provided the mathematical language for the newly born science.

As the contours of this new landscape evolved, a new branch of mathematics was born. Gambling had been the incentive to calculate the odds and some simple arithmetic had sufficed. By the middle of the seventeenth century, however, attention shifted from calculating the odds to the mathematics underlying games of chance. We even know exactly where and when this new branch of mathematics saw the light of day: in the middle of the summer of 1654, in the correspondence between two of the most remarkable mathematicians to have ever walked upon French soil.

The Child Prodigy

As a child, Blaise Pascal (1623–62) was not only exceptionally talented, but also very tormented. His older sister Gilberte wrote a biography about his extraordinary life. She tells us that her brother at the age of eighteen hardly knew a day without pain. He suffered from partial paralysis, which would repeatedly force him to walk on crutches; he regularly experienced severe migraine attacks; and, as if that were not enough, he also had severe intestinal problems.

Blaise is only three when his mother dies after giving birth to his sister Jacqueline. His father, Étienne Pascal, is a high-ranking tax officer in the city of Clermont. He is so rich that he actually does not need to work at all. When Blaise is eight, his father decides to retire. He invests his capital in government bonds and the Pascal family moves to Paris.

In those days, Paris is the radiant center of French culture and science. In the famous salons, the nobility and the rich bourgeoisie meet in search of entertainment. Intense intellectual discussions take place at the numerous Academies where scientists and scholars gather. In the bustling capital of France, Pascal's father hopes to provide his gifted children with the best education possible.

Paris and Mersenne's Circle

Father Étienne sees to the upbringing of his children largely on his own, which was quite unique at the time. Blaise and his two sisters spend most of the day reading Greek and Latin writers. Soon, however, Blaise starts asking his father about geometry. Étienne is afraid that his son, fascinated by mathematics, will neglect his lessons in Latin and Greek.

He tells Blaise he has to wait to study geometry until the age of fifteen. But there is no stopping young Pascal. After Étienne discovers that his twelve-year-old son is busy proving Euclid's thirty-second theorem — the theorem that the sum of the angles of a triangle is always 180 degrees — he gives Blaise permission to study Euclid's work.

Blaise is fourteen when his father decides to take his son with him to the academy of Father Mersenne. Marin Mersenne (1588–1648), a clergyman with a heart for science, had gathered some of the most important mathematicians and scientists around him. His learned society is intended as a platform for openly discussing scientific discoveries instead of keeping them to himself. Father Mersenne also takes care of the correspondence between its members and scientists outside France, such as Galileo in Italy and Descartes, who resides in the Netherlands. The news that a mathematical child prodigy is attending Mersenne's meetings spreads like wildfire.

At only sixteen years old, Pascal publishes an essay on conic sections (the figures one gets when a flat surface cuts a cone at an angle). In the short essay Pascal presents a proof of a new geometric theorem that has borne his name ever since. Descartes is convinced that the work has been written by Blaise's father, but Mersenne assures him that it really is the work of Blaise himself. The great French mathematician and philosopher can hardly conceal his envy and reacts with contempt.

Times are Changing

Two years later, however, fortune turns the tables on the Pascals. The family's invested capital has largely vanished, and as result of his public protests against tax increases father Étienne has fallen out of favor at the royal court. The Bastille awaits him.

Fortunately, his daughter Jacqueline, who is an extraordinarily gifted poet, has been a welcome guest at the royal court from an early age. She manages to regain the favor of the mighty cardinal Richelieu. Her father Étienne is made a high official at the tax office in Rouen. To assist his father with the boring and laborious calculations he has to perform every day, Blaise Pascal invents a mechanical calculator, known as the Pascaline. The machine can add and subtract numbers up to six digits.

But then, at the age of 63, father Étienne dies. Blaise and his sister Jacqueline have to fend for themselves. Nine years earlier, Gilberte had married and left home. Soon Jacqueline leaves home as well. She becomes a nun at the Port-Royal monastery.

The monastery is a stronghold of Jansenism, a strict Catholic movement named after the Dutch bishop Cornelius Jansen. The movement had many followers in higher circles in both the Netherlands and France. At the instigation of the pious Blaise, the Pascal family had also embraced the doctrine of Jansenism. Pascal's conversion to Jansenism would have a major impact on his later life.

The Chevalier de Méré

Being all on his own, Pascal starts looking for some company and distraction. He visits the salons and meets with friends. Among them is Artus Gouffier (1627–96), the young Duke of Roannez, whom he has known since early childhood. The duke has a great interest in mathematics. In the autumn of 1653 he takes Pascal with him on a journey to the province of Poitou. They talk about mathematics along the way.

On the trip they are accompanied by two of the duke's friends. One of them is a certain Antoine Gombaud (1607–84) or, as he prefers to call himself, Chevalier de Méré. In the history of probability theory he is usually briefly referred to as a professional gambler, but this doesn't do justice to the stature of the Chevalier de Méré.

The chevalier belonged to the lower nobility — he owned a small castle in the provincial town of Méré — yet he was a prominent courtier at the court of Louis XIV. He served the Sun King as an advisor on delicate matters. Understandably, this made him a welcome guest in the Parisian salons. The Chevalier de Méré embodied the ideal of the *honnête homme*, a gentleman who possessed a general knowledge of the fine arts as well as the sciences, and who was well versed in the art of conversation.

So De Méré was a man of the world who knew the pleasures of life. He was also a poseur who did not shy away from exaggeration. But during the journey to Poitou, the connoisseur of the pleasant life and the pious Pascal, in spite of all their differences, become friends.

The Gambling Frenzy

One of the pleasures of life De Méré knows everything about is gambling. Gambling was probably the most important pastime in the salons as well as at court. At the royal court the male courtiers joined their female company at the beginning of the evening and, if no opera was scheduled, they would spend their time together gambling until well past ten o'clock. They gambled without hesitation and the stakes were sky-high. In the salons in Paris things were no different. Whether Pascal ever participated in this collective frenzy of the rich we don't know.

In 1654, probably at the beginning of the summer, De Méré presents Pascal with two gambling problems. The first is a fairly simple problem. However, De Méré, who in Pascal's eyes is by no means a true mathematician, claims with utter conviction that the solution defies the laws of arithmetic. The second problem is far more interesting, both mathematically and historically, and not so easy to solve.

Two gamblers play a gambling game in which the odds are equally divided. The one who is first to win six games wins the pot. However, at a certain point, the game has to be aborted. There can be no further play. The problem is how to divide the pot in such a way that each player gets a fair share.

De Méré would later boast that his problem had led to new insights in mathematics. However, he had not invented the problem himself. In fact, it was an ancient mathematical problem, dating back to at least the fourteenth century. Yet, so far, no one had come up with a satisfactory solution.

The Problem of the Unfinished Game

The problem of the unfinished game, in mathematics known as *the problem of points*, first appears in two anonymous Italian manuscripts from the fourteenth century. It may even be of Arabic origin. In 1494 it appears for the first time in print in a mathematical work by the Franciscan monk Fra Luca Pacioli (1445–1517), Leonardo da Vinci's math teacher.

Pacioli formulates the problem as follows. A ball game is played between two teams. The teams are well matched. In other words, the odds of winning a game are fifty-fifty. Each game is worth 10 points. The team that first reaches the total score of 60 points will win the match and receive the prize money of 10 ducats. For some reason, the match has to be abandoned at a score of 50 to 20. The prize money will be split between the two teams. What share of the prize money is each team entitled to?

Your first guess might well be that the money should be divided in proportion to the score at the moment the match is abandoned. This is exactly what Luca Pacioli proposed. Each team gets the percentage that corresponds to its score. To calculate these percentages you simply add the points scored by both teams: 50 + 20 = 70. Now the leading team receives 50/70, that is 5/7 of the 10 ducats, whereas the remaining 20/70 or 2/7 of the 10 ducats goes to the other team. A fair deal, right?

Niccolò Tartaglia, Cardano's relentless rival, thought otherwise. Suppose the match had been abandoned at a score of 10 to 0. Then, according to Pacioli's method, the leading team would get all of the prize money and the other team nothing. Would that be fair? Not at all. The match had only just begun and could still have unfolded in many different ways.

Tartaglia comes up with another solution. Each team gets its money back, yet the leading team receives from the opposing team a sum of money that is proportional to its lead. Tartaglia does not have much faith in his own solution. The problem, he writes, should be referred to a legal committee, not to a mathematician.

Tartaglia's doubts are quite understandable, for his solution has drawbacks of its own. Still, he is wrong to think that the problem is a legal one. A mathematical solution that does justice to both teams exists. To see how, one has to approach the problem from a different angle than Pacioli and Tartaglia did.

What counts is not the number of points that each team has obtained, but the number of points each team still lacks in order to win! The chance of winning the match if it were to be continued determines one's fair share of the prize money. In other words, the prize money should be divided in the same proportion as the odds of winning the match. As Cardano's rule states, this ratio would be a fair one.

The first one to approach the problem in this way was Girolamo Cardano himself. Cardano dismissed Pacioli's solution in clear terms. The mistake the monk made was so obvious that even a small child can see it. His solution only leads to absurdities. But Cardano, despite his understanding of the problem, also failed to find the right solution. His answer happens to be close, but his method is wrong. So when De Méré presents his version of the problem to Pascal, it has been waiting for a solution for more than a century.

Presumably, Pascal sets out to solve the problem straight away. To his great joy he succeeds in finding a solution. But where to find someone who can confirm that he is right? De Méré, being a gambler, knows how to calculate some simple odds. However, he is anything but a mathematician. He persistently claims that the number of points on a line is finite, no matter how often Pascal explains to him that a line is divisible into an infinite number of points. Also the prominent French mathematician Roberval turns out to be of no avail. Pascal decides to write a letter to Pierre Fermat, an old friend of his father's.

The Counselor

Unlike Pascal, Pierre Fermat (1607–65) does not spend the greater part of his life in the higher realms of mathematics. Most of the day Fermat is occupied by more earthly matters. He is a counselor at the Royal Court of Appeal in Toulouse, the *Parliament de Toulouse*.

At an early age, Fermat had gone studying civil law at the University of Orléans, one of the most renowned law schools in Europe. He finished his law studies at the age of 21 and became a lawyer in Bordeaux where he joined a circle of enthusiastic mathematicians. The young lawyer intensely studied the ancient Greek mathematicians as well as the works of the famous French mathematician François Viète (1540–1604). It was the beginning of his lifelong passion for mathematics.

Two years later, he moved to Toulouse. His late father, who had been a wealthy wholesaler, had left him a substantial inheritance. For a huge sum of money Fermat bought the post of counselor at the Toulouse Court of Appeal. This was the usual procedure for obtaining an important office, that is, an office that provided access to the aristocratic *Robe de Noblesse*. From then on, Fermat was officially allowed to call himself Pierre *de* Fermat, but he never did. Not long after, he married the sixteen-year-old daughter of a distinguished colleague. At the age of 23, Fermat is settled for life.

Until the end of his life, Fermat will dutifully fulfill his demanding office. He is widely known for his integrity. He once was one of the judges in a trial against a falsely accused priest. Fermat considered the man innocent. For political reasons, however, the poor soul was condemned and hanged. His body was burned and his ashes were scattered to the wind. For months, Fermat was so upset that he was barely able to carry out his work.

However, Fermat distinguishes himself not only as a competent lawyer, he also possesses a vast knowledge of languages. In addition to French, he speaks fluent Italian, Spanish, Latin, and Occitan, the spoken language in the south of France. He writes poems in several languages, including Latin, in which he expresses himself with an admirable elegance. He is also well versed in ancient Greek. People regularly come to consult him about difficult passages in the text of the Greek writers. But most of all, he has an exceptional talent for mathematics.

Fermat's Theorems

When his childhood friend Pierre de Carcavi moves to Paris and joins Mersenne's circle, Fermat receives a letter from Father Mersenne. The letter opens new doors for Fermat. It is the beginning of his correspondence with great scientists all over Europe.

In these years Fermat makes his groundbreaking discoveries in number theory. He finds inspiration in his favorite book, the *Arithmetics* of the ancient Greek mathematician Diophantus. In his letters Fermat puts forward new theorems about whole numbers. But although he always outlines his method of proof, he never provides the proofs themselves!

Maybe he wanted to encourage his mathematical friends to come up with the proofs on their own. Or perhaps he simply lacked the time to write them out in full. After his death only a single proof was found. Yet most of his theorems later turned out to be true. One of Fermat's theorems, however, escaped proof for centuries. It became known as *Fermat's Last Theorem.*

In the margin of his copy of Diophantus's *Arithmetics* Fermat had written that he had an astonishing proof of the theorem but, alas, the margin was too small to write it down in. Over the centuries, thousands of mathematicians and as many non-mathematicians attempted to prove the theorem.

It was not until 1994 that the English mathematician Andrew Wiles provided a proof. He had been working on it in silence for six years. If Fermat had really ever had any proof, it would certainly not have fitted in the margin. The proof that Wiles provided comprised no fewer than 130 pages of advanced mathematics!

Fermat's Silence

In 1644 a deep silence envelops Fermat. For ten years, he writes hardly anything about mathematics. Fermat experiences difficult times. He is forced to make some distant official journeys, and when he finally returns to Toulouse, a civil war has broken out in France. Fermat finds himself in the middle of the conflict. With great effort, he manages to avert the imminent looting of his native town, Beaumont-de-Lomagne, by the king's ruthless troops.

In the autumn of 1652 another major threat emerges. Toulouse experiences an outbreak of bubonic plague. Fermat falls victim to the dreaded disease but luckily escapes death. Severely weakened, he slowly recovers.

Then, at the end of July 1654, his haunted life takes an unexpected turn for the better. A letter arrives from Paris. The letter is sent by an enthusiastic Blaise Pascal. The former child prodigy writes that he has succeeded in finding a solution to the problem of the unfinished game. He asks the learned counselor if he would be so kind as to have a look at it and confirm that his solution is correct.

Presumably inspired by the young mathematician's fervor, Fermat starts writing about mathematics again. The correspondence between the two brilliant mathematicians would become one of the most remarkable in the history of mathematics.

Unfortunately, Pascal's first letter and Fermat's reply have both been lost, as well as some of their other letters. But the ones that survived still provide us with a clear picture of the solutions Pascal and Fermat found for the *problème des partis*, "the problem of distribution", as Pascal called it.

The Problem of Points

The precise formulation of the problem by De Méré is unknown to us. In Pascal's version two players make a bet. The player who is the first to win three times wins the bet and gets the pot. The bet is prematurely broken off at a score of 1 to 0. How can one give each player a fair share of the pot? We will present a variant of our own, one that is not too difficult and that is in keeping with the original version.

Two players play an old ball game, a precursor to modern tennis. The first player to win six games wins the pot. The prize money is 80 gold ducats, a currency that was in use in the sixteenth and seventeenth centuries. Let's give the players some self-explanatory Italian names. We will name player 1 *Primo* and player 2 *Secondo.*

Primo leads by 5–3 when the game has to be abandoned. Whatever the reason — the only ball gets lost, one of the players is injured, it starts to snow heavily — no further play is possible. Still the organizers of the tournament do not want the players to go home empty-handed. But how to divide the prize money between the two players?

We assume that the players are an equal match. So each has the same chance of winning a game. Obviously, a fair distribution should reflect the chances that each player has of winning the match if it were to continue. Less obvious is how these chances should be calculated.

In his first letter to Fermat, Pascal has apparently asked the counselor to come up with a solution of his own. If their answers agree, this will confirm that Pascal's solution is right. On July 28, Pascal receives an answer from Fermat. The next morning he immediately writes an impassionate letter in return: "My impatience is as great as yours. Although I'm still in bed, I can't help but tell you that last night I received your letter from Carcavi about the problem of distribution. [...] I would like to open my heart to you if I could, so glad am I to see that we are in agreement. I am delighted to see that the truth in Toulouse is the same as in Paris!"

Fermat's Method

Knowledge of how to calculate simple odds was probably fairly common by this time, at least among gamblers. One enumerates all the possible outcomes. The ratio between the favorable outcomes and the unfavorable ones will tell you what the odds are.

The problem of the unfinished game is a bit more complicated, though. The above method presupposes that all possible outcomes are equally probable. The possible outcomes of the tennis match, on the contrary, are not. For example, the probability that Primo wins at 6–3 is clearly greater than the probability that Secondo fights back and wins at 5–6.

Suppose the match were to be continued. There are several possible outcomes. Primo wins 6–3; Secondo wins a game but Primo makes it

6–4; Secondo wins two games in a row but Primo makes it 6–5; or Secondo wins three games in a row and wins 5–6.

These are the only four possible outcomes. However, they are not all equally probable. Let P stand for a game won by Primo and S for a game won by Secondo. Then the four possible outcomes, if the match were to be continued, can be denoted as P, S P, S S P and S S S. Clearly some are more likely, others less.

Fermat comes up with a clever way out. First of all, he determines the maximum number of games that would have to be played in order to be sure that one of the two players has won the match. Primo lacks only one game, while Secondo needs three. So the maximum number of games that has to be played is (1 + 3) – 1 = 3. You have to subtract one game since they can't both win the match.

Imagine that the match fictitiously goes on until this maximum number of three games has been reached, even though one of the players may already have won in fewer than three games. This would give you the following eight possible outcomes: P P P, P P S, P S P, S P P, P S S, S P S, S S P and S S S. This time all possible outcomes have the same probability: 1 out of 8.

Fermat then observes that as soon as Primo wins one game, he has won the match, even if the players still have to continue to play. The reason is simple. Secondo needs three games in order to win the match, but if Primo wins one game, there are only two games left to go. After all, the maximum of games to be played is three. Secondo will always be one game short. This means that we only have to count the outcomes in which Primo wins at least one game.

There are seven outcomes in which Primo wins at least one game. Secondo in his turn only wins at S S S. So the odds are 7 to 1 in favor of Primo. Primo is therefore entitled to a part of the prize money seven times as large as the part to which Secondo is entitled. In other words, Primo gets 70 ducats, Secondo the remaining 10 ducats.

Pascal calls Fermat's approach the *méthode des combinaisons*, "the method of combinations". Although it is a clever solution, it has some major drawbacks. For instance, what if the match is abandoned at a score of 4–2? The play should then be continued virtually for the maximum number of (2 + 4) – 1 = 5 games. This means that we no longer have to deal with eight, but with the much larger number of $2 \times 2 \times 2 \times 2 \times 2 = 32$ possible outcomes (in mathematics, this product is usually abbreviated as 2^5, and called the 5th *power* of 2). As Pascal writes to Fermat: "Your method is accurate, and it was the first to come in my mind while I was looking for an answer. But because finding all combinations requires a lot of effort, I have found a faster way to do so, a completely different method that is much shorter and more elegant".

Pascal's Method

Pascal's method is completely different. It is based on the average sum of money you expect to win at different scores of the match. Before discussing Pascal's method, let's first illustrate how to calculate this "average expectation".

Imagine that two people are playing some kind of ball game. The match has almost come to an end. The score is tied and the player whose turn it is has one last chance to untie the match. If he scores, he wins the full amount of the prize money, 1200 euros. If he doesn't score, the match ends in a draw and both players share the prize money, with each getting 600 euros.

The scoring percentage of the player is one in two. Half the time he will score, half the time he won't. Based on this scoring percentage the average sum of money you expect the player to win is equal to $1/2 \times 1200 + 1/2 \times 600 = 900$ euros. In mathematics this is called the *mathematical expectation* or also the *expected value*. It is one of the fundamental concepts in modern probability theory and Pascal is the first to use this concept.

The prize money Primo and Secondo are each entitled to, Pascal states, is the average amount that each would be expected to win if the match were to be continued. The easiest way to calculate this average is by means of a *tree diagram*. This is a schematic representation of all possible scenarios and their corresponding probabilities. This is what it looks like:

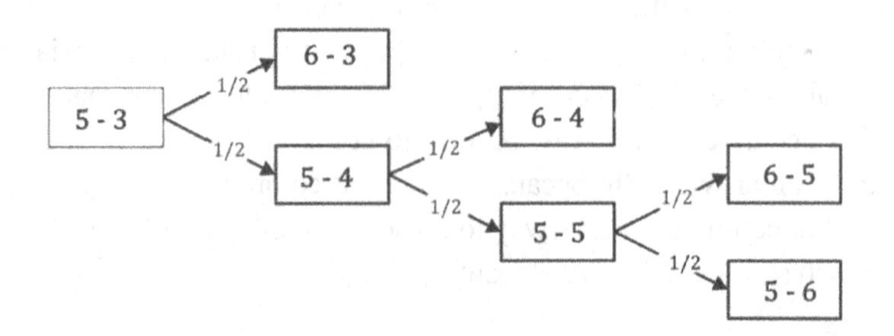

At the time Pascal wrote his letters to Fermat, the concept of mathematical probability still didn't exist (nor did tree diagrams for that matter). However, for the sake of convenience, we will use the concept and reformulate Pascal's solution in a more modern way, using our example of a tennis match as an illustration.

Pascal proceeds as follows. What if the score had been 5–4 instead of 5–3? In that case Primo would have a chance of 1/2 to win the total prize money of 80 ducats. After all, he only has to win one game (6–4).

Alternatively, if Primo loses the game, the score would be 5–5 and both players would each have a chance of 1/2 to win the match. In this case, the expected value for each is half of the prize money, that is, 40 ducats.

So at a score of 5–4, the average amount that Primo is expected to win is $1/2 \times 80 + 1/2 \times 40 = 60$ ducats.

We are now able to calculate Primo's share of the prize money also at a score of 5–3. Either Primo wins 6–3 and is entitled to the total amount of 80 ducats, or he loses and the score is 5–4. But then, as we have just calculated, Primo is entitled to 60 ducats. So the average amount of money that Primo is expected to win at score 5–3, is $1/2 \times 80 + 1/2 \times 60 = 70$ ducats. Secondo gets the remaining 10 ducats. In Pascal's own words, the truth in Toulouse is the same as in Paris!

Fermat's clever method of combinations and Pascal's elegant recursive method betray a mathematical ease you might find easier to admire than to imitate. However, there is no need to despair. The problem can also easily be solved by means of three simple rules of probability: the *product rule*, the *sum rule*, and the *complement rule*.

The Product Rule

Let's take a look at the tree diagram. There are three possible scenarios by which Primo will win the match. The probability that he wins 6–3 is quite obvious: it's 1/2. How about the probability that Primo wins 6–4? Secondo would have to win a game first and then Primo would have to win the next game. So the probability that Primo will win 6–4 is $1/2 \times 1/2 = 1/4$. And how about the probability that Primo wins by a score of 6–5? Secondo would have to win two games in a row first, then Primo would have to win the final game. The probability of Secondo winning two games in a row is $1/2 \times 1/2$. Next, Primo has a chance of 1/2 to win 6–5. So the probability that Primo will win 6–5 is $1/2 \times 1/2 \times 1/2 = 1/8$.

In other words, the probability of each individual scenario of winning the match can be found simply by multiplying the probabilities along the corresponding branches of the tree diagram. The principle behind this method is called the *product rule*. This rule states that the probability that two events both occur is equal to the product of the probabilities of each individual event. The same rule also applies to more than two events.

The only requirement is that the events are *independent* of one another. Two events are independent if the occurrence of one of them does not affect the probability of the other occurring. In our instance, regardless of the outcome of the previous game, the chance of winning a game is always 1/2. So the individual events (i.e. winning a game) are independent of one another. Just like when you toss a coin.

The Sum Rule

We have calculated the probabilities of the separate outcomes of the match. The probability of Primo winning 6–3 is 1/2, the probability of winning 6–4 is 1/4, and the probability of winning 6–5 is 1/8. But what we want to know is the overall probability of Primo winning the match. This probability is simply the sum of the probabilities of the three winning scenarios: $1/2 + 1/4 + 1/8 = 7/8$. Exactly the same as Fermat and Pascal calculated!

The rule that we have just applied is called the *sum rule*. The sum rule is a very intuitive rule of probability. It states that if two possible events are mutually exclusive, the probability that either one or the other will be realized is equal to the sum of their individual probabilities. Again, the same rule also applies to more than two events.

Mutually exclusive events are events that can't happen at the same time. Obviously, our case meets this requirement. It is impossible to win with different scores — for instance, with both 6–3 and 6–5 — at the same time. The following example illustrates what happens if you do add up the probabilities of events that are not mutually exclusive.

Lotteries, as the popular one-liner goes, are a tax for people who are bad at math. Now, imagine a city where 25% of the inhabitants are mathematicians and 75% are loyal participants in the local lottery. It obviously doesn't follow that the probability of a randomly chosen resident being either a mathematician or a participant in the lottery is equal to 75% + 25% = 100%. Clearly, the two are not mutually exclusive. Some inhabitants may not participate in the lottery even though they have no understanding of mathematics. Conversely, some mathematicians may secretly (or not) participate in the lottery.

So if you simply add both percentages you run the risk of counting the people who are not only mathematicians but also participate in the lottery twice, as happens in the following instance.

Suppose you blindly draw a card from a stack of 52 cards. What is the chance that this card is a heart or an ace? There are 13 hearts, so the chance that it is a card of hearts is 13/52. There are 4 aces, so the chance that it is an ace is 4/52. If you add up these probabilities you get a chance of 17/52, but then you are counting the ace of hearts twice! The chance of drawing the ace of hearts is 1/52. So the chance we are looking for is 17/52 − 1/52 = 16/52, that is 4/13.

The Complement Rule

Let's take a second look at the tree diagram. You may notice there is an even easier way to calculate Primo's chance of winning the match. Why not calculate the chance that Secondo will eventually win instead of Primo?

Secondo has to win three games in a row. According to the product rule the probability of winning three games in a row is $1/2 \times 1/2 \times 1/2 = 1/8$. Since either Primo or Secondo will win the match, Primo's chance of winning the match is simply $1 - 1/8 = 7/8$. Exactly the same result as we calculated before.

The probability that an event does *not* occur is called its complementary probability. For this reason, the above method is known as the *complement rule*. The rule often serves as a kind of "magic wand". Many seemingly laborious problems can be easily solved by using this rule, as the following illustration shows.

Suppose you bet that in four throws of a die you will throw a six at least once. What are the chances of winning this bet? One could of course systematically enumerate all the possible outcomes and count the favorable ones, but that would be a lot of work. It is much easier to calculate the complementary probability, the probability that you won't throw a single 6 in four throws.

The probability of not throwing a 6 in one roll is 5/6. So the probability of not throwing a 6 in four consecutive throws is $5/6 \times 5/6 \times 5/6 \times 5/6 = 625/1296$. The probability we are looking for is therefore $1 - 625/1296 = 671/1296$. In other words, the odds are 671 to 625 in your favor.

These three rules of probability not only enable us to solve the above instance of the problem of points, but will also be of great help to unravel the counter-intuitive logic of chance in everyday life. They form the basic toolkit that we will need in Part Two.

Pascal's Formula

Let us return to Pascal's elegant method of solving the problem of points. The same recursive principle can also be applied to solve the example given by Fra Luca Pacioli. For now that we have solved the problem at a score of 5–3, it is quite easy to calculate the odds when the match is abandoned at 5–2. A quick glance at the problem will show you how.

However, if the number of games left to be played is much higher, even Pascal's method becomes quite laborious. Pascal would not have been a true mathematician if he had been content with solving only some individual cases of the problem. So he searched for a general formula, a formula for every conceivable number of games that are left to be played. In a quite surprising way, Pascal actually found a formula that would solve — as he proved — every possible case.

We will not go into the details of this formula and its mathematical proof. After all, that would be contrary to the spirit of this book. Yet, obviously, we can't ignore that Pascal was the first to go beyond simply calculating the odds and explore the mathematics behind games of chance. To give you some idea of this major step, we will explain in general terms how Pascal arrived at his formula. Never mind if you do not understand everything right away. The important thing is to get a clear idea of Pascal's historical significance in the emergence of what he himself called the mathematics of chance.

Pascal's Triangle

When De Méré presented Pascal with the problem of the unfinished game, Pascal was composing a mathematical treatise on a special number triangle that would become known as *Pascal's triangle*. A Latin version of the treatise was probably already finished at the time De Méré presented his two problems.

Not much later Pascal must have decided to write a new version, this time in French: the famous *Treatise on the Arithmetic Triangle* (*Traité du triangle arithmétique*). It contains an extra chapter not found in the Latin version exclusively devoted to the problem of the unfinished game. At first glance a strange addition to a treatise on a number triangle, but its raison d'être will soon become clear. The treatise was not published until three years after Pascal's death, but we know from their correspondence that Pascal had already sent a copy to Fermat.

The figure below shows how Pascal himself drew the arithmetical triangle. The arithmetical triangle is one of the most beautiful and surprising number patterns in all of mathematics. It was already known centuries earlier, but Pascal was the first to systematically explore all of its mathematical properties.

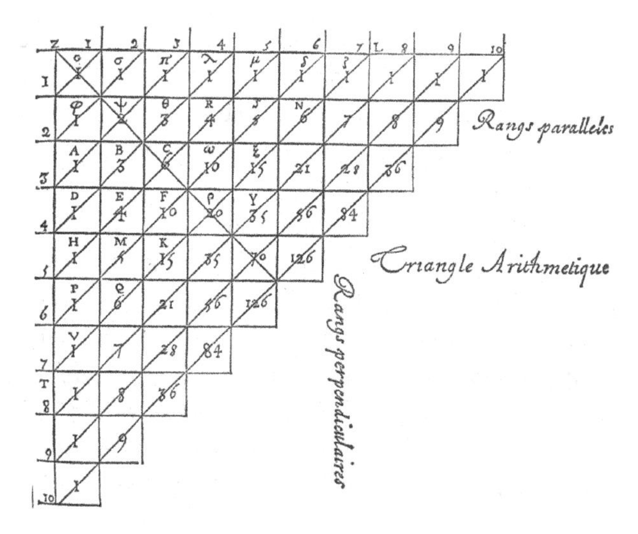

The triangle (see below) is composed in the following manner: the top of the triangle is a 1 and each row begins and ends with a 1. Every other number is the sum of the two numbers directly left and right of it in the row above. Evidently, the triangle can be continued infinitely.

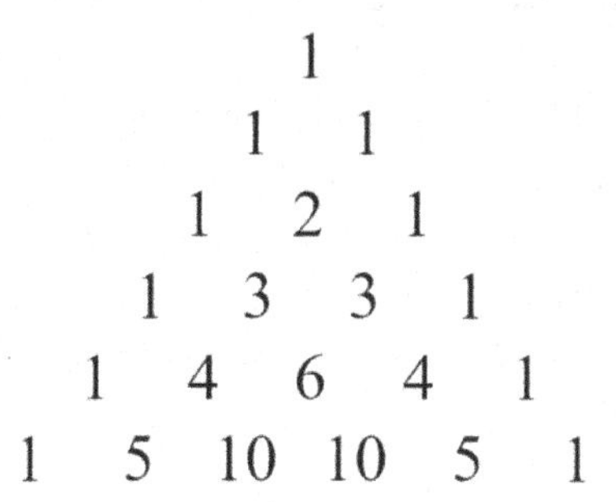

There are lots of interesting things to discover in Pascal's triangle. For instance, each row is constructed symmetrically. It reads the same from left to right as from right to left. The diagonal 1 2 3 4 5 ... indicates exactly the rank of each row of numbers (if we consider the 1 at the top to be row "zero"). Furthermore, the sum of the numbers in each row is equal to a power of two: the "zero" row $2^0 = 1$, the first row 2^1 (= 2), the second row 2^2 (= 2 × 2), and so on. In the diagonals of the triangle several kinds of number series are hidden.

When Pascal did some puzzling with various instances of the problem of the unfinished game, he made a very surprising discovery. He observed that the different outcomes were related to the numbers in the arithmetic triangle! The easiest way for us to understand this remarkable finding is by means of Fermat's method.

Fermat's trick was to imagine that the match would be continued until a maximal number of games had been reached: the sum of the games each player requires to win the match, minus one.

Now it is easy to see that the actual course of the match doesn't matter. It is only the *number* of games won by each player that counts. For as soon as one of the players has succeeded in winning the match, the number of games that is still left to be played will always be one game short of the number of games that the other player needs to win the match.

Consider the initial example of a match abandoned at the score of 5–3. Primo requires one more game in order to win, Secondo three. We should therefore imagine the match to be continued for another three games. If Primo wins one, two, or three times, he will win the match. Only if Primo loses all three games will Secondo be victorious.

If Primo wins three games, {P, P, P}, this can only be done in one way, namely P P P. But if Primo wins two games, {P, P, S}, then there are three possible ways: P P S, P S P, and S P P. Similarly, there are also three possible ways if Primo wins only one game {P, S, S}: P S S, S P S, S S P. And if Primo doesn't win any game at all, {S, S, S}, there's only one way to do so: S S S. This is what it looks like if we put these numbers in a schedule:

Primo wins	{P, P, P}	1
Primo wins	{P, P, S}	3
Primo wins	{P, S, S}	3
Secondo wins	{S, S, S}	1

As you see, these numbers are exactly the same as the ones in the third row of Pascal's triangle! If you add up the numbers for each player, you will arrive at a division of 7:1. And this is exactly the same ratio we calculated before. Coincidence? Let's see what the scheme would look like when the match is abandoned at 5–2, as in Fra Luca Pacioli's example:

Primo wins	{P, P, P, P}	1
Primo wins	{P, P, P, S}	4
Primo wins	{P, P, S, S}	6
Primo wins	{P, S, S, S}	4
Secondo wins	{S, S, S, S}	1

Again, we encounter the numbers from Pascal's triangle, this time the numbers in the fourth row: 1 4 6 4 1. If you add up the numbers for each player, you arrive at a distribution of 15:1. Exactly the same solution Pascal's recursive method yields.

Clearly, there is a pattern here. Let's go back to the initial example. The score is 5–3 and the first player to win six games wins the pot. Fermat's method requires *three* more imaginary games to be played. So we should take the *third* row in Pascal's triangle (1 3 3 1).

If player 2 (Secondo) requires three more games in order to win, you simply add up the first three numbers in the row. This is the number of possible ways player 1 (Primo) can win the match: 1 + 3 + 3 = 7. This makes sense, since the more games the opponent requires in order to win the game, the more favorable the odds are for the other player.

The remaining number, the one game that player 1 (Primo) requires in order to win, equals the number of possible ways for player 2 (Secondo) to win the match: 1. So the odds are 7 to 1 in favor of Primo.

Let's consider one more example. Imagine the same match as the one above is played, but this time it is abandoned at 4–3. Now the maximum number of imaginary games that needs to be played to finish the match is (2 + 3) – 1 = 4. This means we should get to the *fourth* row in Pascal's triangle: 1 4 6 4 1.

Player 2 has to win *three* games to win the match. So the number of possible ways for player 1 to win the match would be the sum of the first *three* numbers: 1 + 4 + 6 = 11. The number of possible ways for player 2 to win the match would be equal to the sum of the last two numbers: 4 + 1 = 5. So the odds are 11 to 5 in favor of Player 1. Player 1 will therefore get 55 ducats, while player 2 will get the remaining 25 ducats. Both

Fermat's method of combinations and Pascal's recursive method will yield exactly the same result.

Pascal offers in his treatise rigorous proof that the above is always the case. The proof is based on his recursive principle (which he calls his *règle des partis*) and a few mathematical properties of the arithmetical triangle. For every unfinished match between two players the arithmetic triangle gives the odds, and consequently a fair distribution of the prize money, as long as the two opponents are an equal match. For the first time in history the calculus of the odds obtained a mathematical foundation.

The Numbers in Pascal's Triangle

Pascal's discovery may at first glance come as quite a surprise. How can the problem of the unfinished game be related to an arithmetical triangle? A closer look at the numbers in Pascal's triangle reveals some of the mathematical logic behind this surprising connection. It turns out that the numbers in Pascal's triangle are a special kind of numbers known as *combinatorial numbers*. The best way to understand what these are is by means of an example.

Suppose that three people have to be chosen from the Supervisory Board of a particular company. The Board consists of ten members. Suppose the three are chosen by drawing lots. How many different combinations of three members can be formed in this way?

The answer to this question is pretty straightforward. For the first person to be drawn, there are ten possibilities. This leaves us with a pool of nine board members, that is, possibilities, for the second drawing. And, according to the same reasoning, eight possibilities will be left for the third person to be drawn. The number of possible drawings is therefore $10 \times 9 \times 8$.

Obviously, the sequence in which the three members of the Board are drawn is irrelevant to the composition of the threesome. It makes no difference whether, for example, Cynthia or Michel or Irene is drawn first, second, or last of the three. So we have to divide our answer by the number of possible ordered combinations of three people.

This number is calculated in exactly the same way: $3 \times 2 \times 1 = 6$. So we have to divide $10 \times 9 \times 8$ by 6 to obtain the number of possible threesomes out of ten: $(10 \times 9 \times 8)/(3 \times 2 \times 1) = 120$. A surprisingly high number, given that there are only ten members from which to choose!

Now suppose that you have to choose respectively 0, 1, 2, 3, and 4 people out of a board of 4 members. Then, using the above method, you'll find the answer is exactly the fourth row in Pascal's triangle! Coincidence? No. Take a board of five members from which respectively 0, 1, 2, 3, 4, 5 people are to be chosen. This time, as you can check yourself, you will get the fifth row of Pascal's number triangle: 1 5 10 10 5 1.

In mathematics, these numbers are appropriately called *combinatorial numbers*. We will refer to them using the letter C. For example, C(5, 3) is the number of ways in which you can choose *three* people or objects from a number of *five*. According to the method we explained above, this is equal to $(5 \times 4 \times 3)/(3 \times 2 \times 1) = 10$. And this is exactly the number found on the *third* place of the *fifth* row of Pascal's triangle. In fact, the numbers in the triangle are exactly the combinatorial numbers.

Let us reformulate this result in terms of Fermat's approach. The maximum number of virtual games needed to determine a winner corresponds to the rank of the row in Pascal's triangle. The position in the row corresponds to the number of games in which the match is decided and the combinatorial number on that position gives the number of possible ways of doing so.

As a result of this correspondence Pascal was able to formulate a general formula for solving the problem of points in terms of combinatorial numbers. A formula that enables us to calculate the odds for every imaginable unfinished match between two players, as long as the two opponents are an equal match.

So the reason that Pascal started to write a new version of his treatise on the arithmetical triangle was to prove that the triangle incorporated the solution to the problem of the unfinished game. Yet Pascal had still greater plans.

The Mathematics of Chance

Following his discoveries, Pascal writes an enthusiastic letter in Latin to the *very famous Paris Academy of Mathematics*, as he calls the circle of the late Father Mersenne. In the letter Pascal informs the learned company about the several scientific treatises he is composing or intends to compose. One of his projects stands out from all the others. He intends to write an entirely novel treatise on a wholly new branch of mathematics:

"A completely new treatise on a subject still completely uninvestigated, namely the arrangement of chance in games that obey chance. What in our French language is called *faire les partis des jeux*. That is to say, when fickle chance is contained by the balanced mind in such a way that each of the two players always gets exactly a fair share. [...] For with the help of mathematics we have made her a science with such a degree of certainty that she, having become a part of this mathematical certainty, is already audaciously making progress. And thus, by combining the uncertainty of chance with the force of mathematical proof and by the reconciliation of two apparent opposites, she derives her name from both of them and rightfully assumes the wonderful name of Mathematics of chance!"

The news that Pascal and Fermat had discovered a new branch of mathematics, the mathematics of chance, spread rapidly far beyond the borders of the country. Pascal's mathematical treatise on chance, however, would never see the light of day. For, at the end of the year, his life took an unexpected turn.

Pascal's Mystical Experience

On the evening of November 23, 1654, at half past eleven, Pascal falls into a trance that lasts for two hours. The pious Pascal experiences a kind of mystical revelation. He can only stammer words and short sentences. "Joy, joy, tears of joy!" The two hours he spends in overwhelming bliss mark a turning point in his life. From now on, he will devote his entire life to his faith. He decides not to marry, which was his firm intention in previous years, and renounces the worldly pleasures of his friends.

He even bids farewell to the world of his beloved mathematics. But not completely, for occasionally he will spend some time on solving mathematical problems. Yet from now on his faith means everything to him. He meditates frequently and spends part of his days reading the Bible. He regularly goes into retreat at the Port-Royal monastery.

It would, however, be wrong to think of Pascal as having become a kind of religious hermit. He still keeps in touch with his friends and lives in a modest house in Paris. He invents a reading method for children, meant to be used in Port-Royal's monastery schools.

In 1662, with the help of some of his friends, he launches a large-scale public project meant especially for the less fortunate: *les carrosses à cinq sous*, the world's first public urban transport. The carriages, accommodating six people, transport the inhabitants from one district to another at a modest price according to a fixed timetable. The first coach was inaugurated in March 1662 to overwhelming interest from the population. It would be his last act of charity.

Two years earlier, in 1660, Pascal's health has seriously deteriorated, making him barely able to devote any time to reading or writing. In the middle of the summer of 1662, he still witnesses the launch of the last of the five scheduled services, but his illness gets worse and worse. Although the doctors try to reassure him, Pascal fears for his life and draws up his will. In the early night of August 19, Pascal suffers severe convulsions. Eventually, two months after his 39th birthday, his tormented body gives up the ghost.

His remarkable public project, due to price increases and political provisions that excluded the lowest classes, was also not long-lived. It would not be until the nineteenth century that public transport would be reintroduced into cities such as London and Paris.

Three years after Pascal's death, Pierre Fermat also died, at the age of 57, physically exhausted and weakened by the plague.

Father of the Theory of Probability

The mathematical dexterity displayed by both Pascal and Fermat reveals how the mathematical landscape had changed since Cardano made his first difficult steps into the direction of a mathematics of chance. The time had come for some serious progress. Yet it is remarkable for a branch of mathematics to have such a clear date of birth. It begs the question of who the real father of this newborn child was.

De Méré was very clear about this: he himself. In a letter to Pascal he wrote: "As you know, I have discovered such rare things in mathematics that even the most learned among the ancients have never discussed them. They amazed the best mathematicians in Europe. You have written about my discoveries, as have Mr. Huygens, Mr. de Fermat and many others who were full of admiration".

If we ignore the preposterous egomania of the Chevalier de Méré, which was only surpassed by his unbridled fantasy, we must at least give him the credit for the fact that "his" problem was the catalyst that launched a new branch of mathematics. As regards his posturing as the inventor of the theory of probability, we would do well to respond in the way Leibniz did. With a smile.

But what about Fermat's contribution? Nowadays, Fermat is considered to be one of the greatest mathematicians, if not the greatest, of his age. Aside from other groundbreaking contributions, he is known as the founder of modern number theory. In his correspondence with Pascal, in addition to his method of combinations he alludes to another ingenious universal solution to the problem of the unfinished game. Regrettably, nothing more than this allusion has come down to us.

It was Pascal who, more than Fermat, elevated the "calculus of odds" from the level of simple calculations to the level of real mathematics. He did so by formulating and proving a universal solution to the problem of the unfinished game. His contributions to the theory of probability have had a major impact on its later development. Pascal was the first to use the concept of *expected value*. His recursive method would be widely used by later mathematicians in solving other difficult problems. Finally, combinatorial numbers would play an important role in calculating all kinds of probabilities. For all these reasons Pascal deserves to be called the intellectual father of this new branch of mathematics.

But even if we ignore all these mathematical contributions and only look at Pascal's enthusiasm, inspiration, and vision, we have to agree that probability theory was Pascal's brainchild. Fermat was his experienced sparring partner, perhaps the best he could ever have wished for. In a letter to Fermat Pascal writes that he considers his older friend to be the greatest mathematician in all of Europe. Fermat, in turn, had a very high opinion of the younger Pascal. In a letter to Carcavi he wrote that there was no problem for which Pascal, if he so desired, could not find a solution.

Eternal Fame

The passing of Blaise Pascal did not mark the end of his intellectual contributions. After his death, piles of notes and fragments were found, some of them bundled together. Pascal had intended to write a book to defend the Christian life against the views of some of his liberal friends, but never completed it. In 1670 the collected fragments and one-liners were published under the title *Pensées* (*Thoughts*).

The book became an enormous success. It has never been out of print since, has been published, printed, and translated countless times up to the present day, and is considered to be one of the classics of French literature. The success of the book is largely due to Pascal's thoughts on the uncertain conditions under which mortal humans have to live. On top of this, Pascal was also an exceptionally good writer. His natural and lively prose style would serve as an example for many generations to come, including great writers such as Voltaire and Jean-Jacques Rousseau.

Numerous books have been written about Pascal as a Christian thinker. The former mathematical child prodigy is world famous as one of the greatest writers of French literature. As the intellectual father of probability theory or, as he himself called it, the mathematics of chance, he is scarcely known.

The Dutch Archimedes

The Early Years

In the middle of the summer of 1655 a young man from Holland, still unknown but soon to become famous, arrives in Paris. He is in the company of a cousin and one of his brothers. They sailed to the French coast at the beginning of July and from there continued their journey overland all the way to the French capital, where they plan to stay for the remainder of the summer. Their visit is meant to put the finishing touch on the young man's education. His father, secretary to the Prince of Orange, thought his son would undoubtedly learn the fine art of conversation in the flourishing salons and learned societies of Paris.

At home, the young Dutchman received a private education in all the arts a true courtier should master. Together with his sister and four brothers he not only learned French and Italian but also Greek and Latin, which he already spoke fluently at the age of nine. The children were taught music as well, and within three years he mastered the harpsichord, the lute, and the viola da gamba. Riding and fencing would follow later.

In everything his mind or body touched, the inquisitive son outshone his brothers and sister. But most of all, the young all-rounder proved to have an exceptional talent for mathematics and physics. He was fifteen when he took his first private lessons in mathematics, with exceptional success, as his father wrote. He proudly called his son "my Archimedes", and indeed, the young Christiaan Huygens (1629–95) would grow into the Archimedes of his time.

Between Heaven and Hell

The contributions that Christaan Huygens made to science are too numerous to mention. With the help of his telescope, he discovers the moon Titan near Saturn and comes to realize that the strange bulges on either side of Saturn are actually a flat ring around the planet. He invents the pendulum clock, making it possible to measure time much more precisely than ever before. And at the age of thirty-one he is already so famous as to be granted an audience by the French King Louis XIV.

At the same age, however, he is struck for the first time by the disease that will persecute him for the rest of his life. Christiaan feels exhausted and is incapable of getting out of bed. Servants have to carry him. He barely eats and he no longer sleeps. The king's physician is called in, to no avail. Eventually, the family doctor diagnoses the disease: "hypochondriac melancholy" or, as we would now say, a severe form of depression. It will take him two years to recover and resume his scientific work.

At the age of thirty-seven Christiaan Huygens is entrusted with the scientific leadership of the *Académie Royale des Sciences*, the prestigious scientific institute recently founded in Paris. Among other things, he will work on his famous theory of the wave propagation of light and a 31-tone system of music which makes it easier to change key on keyboards than in the traditional twelve-tone system.

But in 1681, at the age of 52, the merciless disease strikes again. Christiaan seeks refuge with his father, Constantijn Huygens, in his hometown of The Hague. Because of the anti-Protestant sentiments in France he will never again return to Paris, where his presence is no longer appreciated. He will spend most of the rest of his life in the family's mansion, living on an annual allowance from his father.

In 1687 his father dies, leaving his son inconsolable. Christiaan begins to write a Latin treatise entitled *Cosmotheoros* on the existence of life on the other planets. The book will be published in 1698 and ultimately translated into Dutch, English, French, German, and Swedish. Even a Russian translation will be published at the instigation of the Russian Tsar Peter the Great. Sadly, Christiaan will not live to see the success of his book.

In a preliminary study for the book, undoubtedly referring to himself, he comments: "Just as those who have visited on distant journeys many countries and their peoples, judge their homeland more wisely and better than those who have never set foot outside it, so it is with someone who is accustomed to dwelling in his thoughts among the stars, and from there contemplates this little ball of our Earth: he often envisions how small a particle our planet is of the Universe, and imagines at the same time what goes on elsewhere on so many thousands of planets like the earth. How tiny are our kingdoms, what a hassle, what a rush!"

But less than a year after completing the book, his wondrous mind slips away into a hopeless world of darkness. The doctors know no better than to prescribe goat's milk and advise him to take regular baths. He no longer sleeps and fears losing his mind. Thoughts of death torment him. He refuses to receive a preacher and is convinced that people would rend him to pieces if they only knew his views on religion. He hears voices and loses considerable weight from scarcely eating, afraid that his food has been poisoned.

On July 7, 1695, a minister calls, but is sent away. The same night his condition worsens. Christiaan loses consciousness and the next morning he awakes no more. One of the greatest scientists of the seventeenth century has died.

Happy Times

In the summer of 1655, when young Christiaan arrives in Paris for the first time, he is just 26 years of age and still virtually unknown. It will be the happiest time of his life. He has studied law in Leiden and Breda, according to his father's wishes, but he has never taken any exams. Instead he has been following courses in mathematics.

Christiaan immediately feels at home in France's vibrant center of culture and science. The three young men pretend to be knights from Holland, wearing swords and spending money like crazy. Wherever they go, they meet new people, from courtiers to musicians, from writers to scientists.

At Father Marin Mersenne's academy Huygens meets the mathematicians Gilles Roberval and Claude Mylon. It is from them that he presumably first hears about the problem of the unfinished game and about the new mathematics of chance discovered by Blaise Pascal and Pierre Fermat. Yet neither Roberval nor Mylon can provide him with the mathematical details.

Much to his regret, Huygens, who must have been fascinated immediately, does not succeed in meeting the protagonists themselves. Fermat remains in Toulouse and Pascal has retired to the monastery of Port-Royal. Halfway through December, Huygens and his companions return to Holland. Back home, Christiaan writes that if he had not been assured that Pascal had said farewell to mathematics, he would have spared no effort to meet him in Paris.

On the Mathematics of Games of Chance

Christiaan promptly begins to write a treatise in Dutch entitled *Van Rekeningh in Spelen van Geluck* (*On the Calculus of Games of Chance*). In the foreword to the Dutch edition he writes: "One should know that quite recently some of the most famous mathematicians of all France already devoted themselves to this kind of mathematics, so that no one grants me the honor of its first invention, which is not mine. Yet they, being used to test each other with many intricate problems, have kept their solutions secret. As a result, I had to figure out everything from scratch and all by myself …".

In no time Christiaan succeeds in unraveling the problem of the unfinished game, and in the spring of 1656 he sends the Dutch manuscript of his treatise to his former teacher of mathematics, Frans van Schooten. Christiaan Huygens has written his treatise in Dutch, because due to the novelty of the subject, as he says himself, he does not have all the necessary Latin terms at his disposal. Van Schooten translates the text into Latin, and in August or September 1657 Huygens' treatise is published together with Van Schooten's own *Mathematical Exercises*. In 1660 the original Dutch version is published as well.

Huygens' treatise is the first ever published on the new mathematics of chance. Pascal's treatise on the arithmetic triangle will not be published until 1665, and the correspondence between Pascal and Fermat not until 1679. For about half a century Huygens's treatise remains the foremost book on the mathematics of games of chance and is translated into English twice before the end of the century.

The Expected Value of a Game

The treatise, which covers only thirty-two pages, consists of fourteen "propositions". Nine of them are related to the problem of the unfinished game. To the fourteen propositions five more problems are added for the reader to solve by himself.

Christiaan observes that in order to calculate a fair distribution of the prize money, one first has to determine how to mathematically express the value of the games that are yet to be played. The principle that Huygens uses is that of the *expected value*. It is treated in the first three propositions.

Christiaan Huygens is not the first to use the concept of the expected value. Pascal did so before him in his solution to the problem. Yet Huygens is the first to formulate the concept in a universal and systematic way.

Nowadays, we define the expected value by means of probabilities. Suppose you play a game of chance where you win either amount A or amount B. The expected value is then amount A multiplied by the probability of winning amount A, plus amount B multiplied by the probability of winning amount B. If you lose amount B, you simply subtract amount B multiplied by the probability of losing amount B. Intuitively, the outcome — that is, the expected value — is the amount you win or lose on average if you play the game often enough.

Take, for instance, the game of roulette. The ball can fall on the numbers 0 through 36. Of the numbers 1 through 36, half are red and the other half are black. Zero is green. If you bet on red, your chance is therefore 18/37, and the same applies to black. If you win, you are paid an amount of money equal to your bet; if you do not, you are paid nothing and instead lose your bet.

Say you bet 74 dollars on red. Now the expected value equals 18/37 (the probability that you will win) multiplied by 74 dollars minus 19/37 (the probability that you will lose) multiplied by 74 dollars. This amounts to $-1/37 \times 74 = -2$ dollars. So on average you lose 1/37 of your bet. The same holds true for all other possible bets on the roulette game. There is no winning strategy. In the long run the casino always wins.

The Problem of the Unfinished Game

The propositions four to nine, deal with the problem of the unfinished game, or as it is often called, the problem of points. In contrast to Pascal, Huygens does not provide us with a general formula. Each of Huygens' propositions tackles a specific problem with a numeral solution.

Nevertheless, Huygens has arranged the propositions very ingeniously. The solution to each proposition builds on the solution to the previous proposition. This is essentially the same recursiveness that Pascal applies in his method. Thus, starting with a simple two-player problem, Huygens leads the reader all the way to the solution of a complex three-player problem. Since we have already discussed in detail the problem of the unfinished game, we will leave it to the reader to study Huygens' own clear explanation of the problem.

If you read Huygens' treatise, you will frequently come across the word "chance". Yet one should be careful not to ascribe to it our modern concept of probability. The word "chance" goes back to the French *chance*, which itself derives from the Latin word *cadentia* that we encountered earlier in the context of Richard de Fournival's table and which means "the outcome of a die".

In Huygens' treatise, *chance* means various things. Most of the time it means a possible outcome, favorable or unfavorable. When used in the plural, it simply means the odds. It was not until the eighteenth century that the word "chance" finally came to mean the probability of an outcome.

The Titans Challenge One Another

Huygens' treatise is not exclusively devoted to the problem of the unfinished game. When, in April 1656, Huygens sends his manuscript to Frans van Schooten, he writes a letter to Roberval and Mylon about his intention to publish a treatise on games of chance. In the same letter he includes a problem of his own devising. In the treatise it is the very last proposition, number fourteen. Huygens is eager to learn whether the French mathematicians will arrive at the same solution as the one he himself proposes.

Mylon writes back, but unfortunately his solution is wrong; Roberval does not answer at all. Through Mylon and Pierre de Carcavi, however, the problem posed by Huygens also comes to the attention of Fermat and Pascal. The two mathematical giants, whom Huygens to his regret had not been able to meet in Paris, now take up the challenge and both succeed in finding the right solution. What is this problem that Huygens so eagerly wants to be solved?

Imagine you and one of your friends play the following game with two dice. As soon as you throw six pips you win, but if your opponent throws seven pips the victory is his or hers. You take turns throwing the dice. Since the odds are in favor of your friend (6 to 5), you are allowed to start. The question is: which one of you has the greater chance of winning?

If you try to solve the problem in a straightforward way, you will end up with an infinite series of fractions: your chance of winning on the first roll, your chance of winning on the third roll, and so on. Huygens, however, found a shortcut to the solution and is eager to find out if the French mathematicians would also be able to do the same.

Fermat not only sends his solution, but he also replies with five difficult chance-related problems for Huygens to solve. According to Huygens himself, he solves all five problems on the same afternoon that he receives the letter from Fermat — that is to say, he figures out how to solve them, as the "mindless" calculations themselves do not interest him. He adds two of Fermat's problems to the problems that conclude his treatise (problems numbers one and three).

The Gambler's Ruin

Fermat is not the only one to meet the challenge. A few months later in October 1656, out of the blue, Pascal reacts as well. The former child prodigy comes up with a difficult problem of his own. Carcavi writes to Huygens: "Here's another problem that Pascal has put to Mr. Fermat and that he believes to be incomparably more difficult than all the other ones".

Pascal reckons that Fermat will not be able to solve it, but he is mistaken. Almost immediately, Fermat sends the correct answer, but he fails to say *how* he solved it. Huygens also finds the right solution. He adds Pascal's problem as the fifth and final problem at the end of his treatise.

In Huygens' version it reads as follows. Players A and B each have twelve coins and play with three dice under the condition that whenever 11 pips are thrown, A has to give a coin to B, whereas B has to give a coin to A each time 14 pips are thrown. The player who first gets all the coins wins the game. In other words, it is all or nothing.

Finding the right solution is far from easy, yet the solution itself can be expressed in a fairly simple way. To calculate each player's chance of winning, all one needs to know is how many ways the sums of 11 and 14 can be thrown with three dice, and then raise these numbers to the twelfth (the number of coins that each possesses) power.

The sum of 11 can be thrown in 27 ways, while the sum of 14 can be thrown in 15 ways, as you can check for yourself. According to Huygens' formula, the odds are therefore in the ratio of 27^{12} to 15^{12}, or in its reduced form 9^{12} to 5^{12}, which is 282,429,536,481 to 244,140,625 when written out in full. Formulated in percentages, the probability that player B wins is 99.91% and the probability that A wins is only 0.09%! So player A is almost certainly heading for "bankruptcy".

As you will recall, Galileo was asked to explain why betting on the sum of 10 is more profitable than on the sum of 9 when rolling three dice. The gambling friends of the Grand Duke of Tuscany claimed that long experience had taught them so. One would think that with such a small difference between the chance of throwing the sums of 9 and 10 — the difference being only 1/108, less than 1% — even long experience would not have taught the gamblers that betting on 10 is more profitable than 9. Yet if the gamblers had played Pascal's game, the opposite would have turned out to be true.

Suppose they had been playing Pascal's game, each with an amount of twelve ducats. What would their chances of winning be? According to Huygens' formula, the odds are 25^{12}:27^{12}. In terms of percentages, the gambler betting on the sum of 9 has a 28.4% chance of winning and a

71.6% chance of losing. In other words, his chance of going bankrupt is about two and a half times greater than his chance of winning!

A Contemplation of Beauty

Pascal's problem has become a classic one, known in the theory of probability as the *gambler's ruin*. It is a classic problem because it raises new questions, for instance about the duration of the game. But also because it emerges in various forms outside the world of gambling, such as in the fields of physics and biology, as well as in the world of finance.

Neither Huygens nor Pascal, of course, could have foreseen this. Yet Huygens shares the same expectations as Pascal about the new mathematics of chance. Huygens writes in his foreword: "The more difficult as it seemed beforehand to determine by reason what is uncertain and subject to chance, the greater the admiration will be for the science that can bring this about". According to Huygens, each time one looks more closely into the problems he poses in his treatise one will see that the mathematics of games of chance is not just a playful pastime, but also "a profound contemplation of great beauty".

4 The Birth of Probability

The Art of Thinking

The *Logic of Port-Royal*

Neither Pascal, nor Fermat, nor Huygens ever used the term *probability*. The mathematical concept simply did not yet exist. The new mathematics of chance had its roots in the world of gambling. Gamblers didn't calculate probabilities; they calculated the ratio between favorable and unfavorable outcomes. At best one could say that since they were familiar with the concept of favorable outcomes, they had, in a way, an unarticulated understanding of the concept of probability. However, it would not take long before the concept of mathematical probability was introduced. Not in a mathematical treatise, as one would expect, but in a philosophical handbook!

In 1662, the year of Pascal's death, an anonymous philosophical work on logic was published in France. The title of the voluminous book was Logic or the Art of Thinking (La logique ou l'art de penser). In no time it became an overwhelming success. The author turned out to be the well-known theologian and philosopher Antoine Arnauld (1612–94), who had written the work in collaboration with his colleague Pierre Nicole (1625–95).

Antoine Arnauld, little known today, was in his time one of the leading scholars in France. His contemporaries called him *le grand Arnauld.* He had studied theology at Sorbonne University and, like Pascal, had eventually become a follower of Jansenism. Because of his Jansenism Arnauld was expelled from the Sorbonne in 1656. The scholar took refuge in the monastery of *Port-Royal des Champs* near Versailles and became one of its most prominent residents. For this reason his book became widely known as the *Port-Royal Logic*.

The book has been reprinted no fewer than sixty-three times and has been translated into many languages. The Latin version of the book became compulsory reading at numerous universities throughout Europe. Until well into the nineteenth century the *Port-Royal Logic* was the most widely read work in the field of logic.

LA
LOGIQVE
OV
L'ART DE PENSER:

Contenant, outre les Regles communes, plusieurs obſervations nouvelles propres à former le jugement.

A PARIS,
Chez Charles Savreux, au pied de la Tour de Noſtre-Dame.

M. DC. LXII.
AVEC PRIVILEGE DV ROY.

Lotteries and Lightning

The very last chapter of the *Port-Royal Logic* is dedicated to the question how one can arrive at a rational judgement about something that is unpredictable. In any case, not as most people do. Most people, according to Arnauld, only focus on the personal gain they hope for, or on the danger they fear. But a rational judgement about a chance event must not only be based on one's hope or fear. It must also take into account the *probability of the event*. Arnauld gives two examples taken from everyday life.

Most participants in a lottery picture themselves as the winner of the jackpot of 20,000 crowns. They don't take into account that it is 30,000 times *more probable* that you will just lose your money. The same goes for the widespread fear of being struck by lightning. In reality, no more than 1 in 2 million people will die from lightning strike. There is hardly any violent death that is less common! These small probabilities should put one's hopes or fears in perspective.

Here, for the first time in history, we encounter the idea of a numerical probability of a chance event. To us modern readers it sounds very familiar, but to seventeenth-century readers it was completely new. How were they able to understand this new concept? Nowhere in the book does Arnauld give an explicit definition. One could argue that readers could derive its meaning from the examples themselves. But this still doesn't answer the question why Arnauld uses the term *probability*. Why not call it, for instance, *accidentality*?

Probability and Plausibility

Arnauld had not created the term *probability* himself. By the seventeenth century the word had already had quite a long history, going all the way back to Roman times. The Roman lawyer and statesman Cicero was the first to ever use the word.

Cicero considered knowledge of philosophy to be an essential virtue of the educated Roman. In an attempt to introduce his fellow Romans to Greek philosophy, he wrote several Latin treatises on Greek philosophy. For the Greek philosophical terms he conceived corresponding Latin words.

When he needed a Latin translation for the Greek term *pithanos*, "plausible", Cicero coined the word *probabilis* ("probable"), derived from the Latin verb *probare*, "to approve". In this sense it was and still is used to qualify opinions and judgements. In his book *Academics* Cicero himself gives an illuminating illustration: "Does a wise man, when he is going on board a ship, know for sure that he will reach the destination he has in mind? No one can. Yet if he were to set sail from here to Puteoli, which is four miles away, on a robust ship with a capable steersman and the sea is calm, it would appear plausible (*probabile*) that he will arrive there safe and well" (*Academics* 2 100).

So, in its original sense *probability* refers to the plausibility of a judgement. Probability in this sense has little to do with the mathematical probability of a chance event. So how did the word probability obtain its mathematical meaning? The answer can be found in the *Port-Royal Logic* itself.

A New Logic

The traditional logic that was taught in universities was based on Aristotle's predicate logic and on his theory of the syllogism. As in modern logic, its emphasis was on the logical validity of inferences. In the prologue to the *Port-Royal Logic*, on the other hand, Arnauld argues that human errors are usually not the result of faulty inferences. They are the result of false judgements that lead to false conclusions. He has therefore written a new logic that was meant to "educate our judgement". His new logic provided rules to distinguish true from false judgements. "There are different routes practically everywhere", he writes, "some true, others false. It is up to reason to choose among them". But it was not the shifting intellectual landscape of his time that occupied him most.

The Rebirth of Skepticism

In the second half of the sixteenth century a Latin translation of Sextus Empiricus' *Outlines of Pyrrhonism* had become available in print. In this work the ancient Greek philosopher Sextus Empiricus gave an extensive account of Pyrrhonism, a philosophical school that was named after its founder, the Greek philosopher Pyrrho of Elis, who propagated a radical form of skepticism.

Ancient Skepticism was little known in the West, but this publication would change everything. The radical skepticism of ancient Pyrrhonism was embraced and popularized by the French philosopher Michel de Montaigne (1533–92), whose *Essays* were widely read. The changing intellectual landscape proved to be fertile ground and the beginning of the seventeenth century witnessed the rebirth of ancient Skepticism.

The New Pyrrhonists, as they were called, claimed that one cannot distinguish the true from the false, because everything is open to doubt. They used the standard arguments of their ancient predecessors as ammunition and targeted both the religious doctrines and the new science of their time. Arnauld called their skepticism a "mental defect". His new logic was meant as a "remedy".

But where did Arnauld find the ingredients of his remedy? They don't appear in the typical logic books. Arnauld admits that his new reflections are not completely his own. They are largely borrowed from the books of a "famous philosopher of the century". That philosopher was René Descartes.

The Quest for Certainty

Descartes (1596–1650), widely regarded as the father of modern philosophy, was first of all a mathematician. He greatly contributed to the introduction of analytical geometry, which replaced the ruler-and-compass construction of the ancient Greeks with coordinates and algebraic equations. As he himself reveals, he took delight in mathematics because of its certainty and self-evidence.

About the flourishing skepticism of his time he wrote in a rather sarcastic tone: "Neither must we think that the sect of skeptics is long extinct. It flourishes today as much as ever, and nearly all who think that they have some ability beyond the rest of mankind, finding nothing that satisfies them in the common Philosophy, and seeing no other truth, take refuge in Skepticism". Descartes' famous *Discourse on the Method* was his answer to this trendy skepticism.

In the *Discourse* Descartes uses his well-known method of radical doubt. Not to put everything into question, but to find true statements that withstand the most severe forms of doubt. The most fundamental one was the celebrated *Cogito ergo sum* ("I think, therefore I am"). Descartes called it "so firm and sure that all the most extravagant suppositions of the sceptics were incapable of shaking it".

Like the axioms of mathematics, these "doubt-resistant" statements should be the building blocks on which the edifice of the new scientific knowledge was to be erected. Descartes later boasted that he was the first of all men to overthrow the doubts of the skeptics.

Probability and Certainty

Part Four of the *Port-Royal Logic* is entitled *the Method.* In his discussion of scientific method Arnauld adopts much from Descartes' philosophy. But when it comes to making judgements in the practice of life, Arnauld concedes that Descartes' mathematical certainty is not feasible. What's more, even a practical form of certainty is often not attainable. Instead we should settle for a lesser degree of certainty and embrace the most probable alternative. In other words, the most plausible judgement should point the way.

Arnauld explicitly states that he is not talking about moral judgements or the opinions of the learned, but about judgments concerning human and contingent events. That is, events from daily life that may or may not have happened in the past, or that may or may not happen in the future. The probability of such judgements is based on what Arnauld calls *the circumstances*. That is, anything that counts as evidence.

Judgements about past events are justified by circumstances that follow the event. For example, when a judge weighs the evidence that someone has committed a crime. Judgements about future events are justified by evidence that precedes the event, for example, when a physician assesses the outcome of an illness on the basis of preceding symptoms.

Some future events, however, evade this rule. These events, known as *future accidents*, are the subject of the last chapter of the *Port-Royal Logic*.

The Probability of Chance Events

What exactly are future accidents? *Accident* is a philosophical term that goes back to Aristotle. In Book Five of his *Metaphysics*, the ancient Greek philosopher illustrates the concept of an accident (*sumbebêkos* in Greek) by the famous example of a man who, when digging a hole to plant a tree, finds hidden treasure. In short, a future accident is nothing other than a chance event.

Since future accidents are chance events, there are by definition no preceding circumstances that render judgements about future accidents more or less plausible. After all, chance events are unpredictable. Still, chance events themselves can be more or less certain to happen. One could say that they have some kind of inherent plausibility, so to speak. Arnauld calls it *the probability of the event*.

In other words, the new mathematical concept is called probability by analogy. "Probability" was a familiar concept and was discussed in the preceding chapters. The analogy — people often learn new things through analogy — helps to understand the new mathematical concept.

Unfortunately, in the course of history, this analogy has also proven to be a source of confusion. To keep the two senses of *probability* — the concept of plausibility and the concept of mathematical probability — apart, it helps to realize that probability in its traditional sense refers to *opinions or judgements*, whereas probability in its mathematical sense deals with *chance events*. The first is based on rational justification, the second on mathematical calculation.

Mathematical Probability Goes Viral

The fact that the *Port-Royal Logic* was compulsory reading at many universities throughout Europe must have been the reason that the new mathematical use of the term *probability* gradually became widely accepted.

The first edition of the *Dictionnaire de l'Académie française* (1694) defines *probable* as "what appears to be true; what appears to be based on good reasons". A century later, the fifth edition of the dictionary (1798) adds under the term *probability* that "in mathematics, the name probability theory is used for rules to estimate the probability of winning or losing in games of chance".

The Swiss mathematician Jacob Bernoulli (1655–1705), whom we will meet in the next chapter, formally introduced the term *probability* into mathematics. In his masterpiece, *The Art of Conjecturing* (*Ars Conjectandi*, 1713), he defines probability explicitly as a number between 0 and 1 that expresses the degree of certainty of an event.

There can be little doubt that he adopted the term *probability* from the *Port-Royal Logic*. Bernoulli had great admiration for the scholar Arnauld. Most likely he had studied Arnauld's handbook on logic at university.

The Expected Value as a Guiding Principle

The difference between probability in the traditional sense of the word and probability in the mathematical sense can be seen from their different roles as guiding principles. When it comes to forming a rational opinion the best thing one can do is, in Arnauld's own words, "to embrace the more probable". But in the case of future accidents or, as we would say, chance events, things are more complicated.

To make a rational decision about future accidents one must take into account not only the size of the gain or loss, but also the probability of both. In somewhat cryptic terms Arnauld writes: "all these things must be considered together in a mathematical way". To clarify what he means, he gives the example of a simple game of chance.

Ten people each contribute one crown to the pot. Only one person can win the money at stake, while all the others lose. So each player can either win nine crowns or lose one crown. Yet each player has only one degree of probability to win nine crowns, but nine degrees to lose a crown. Mathematically expressed, the expected value is: $1/10 \times 9\ crowns - 9/10 \times 1\ crown = 0\ crowns$.

When it comes to future accidents one must weigh gains and losses according to their respective mathematical probabilities. In other words, in chance events our guide is not the most probable but what we previously called the expected value.

According to Aristotle, a rational account of chance events was not possible. This is, of course, by definition true. Chance events cannot be predicted in advance. There simply are no preceding circumstances that predict the outcome. But Arnauld shows that, with the help of mathematics, there is still a rational way to deal with the uncertainty of chance events in daily life.

These mathematical reflections were completely new in philosophy. So where did the great Arnauld get this new mathematical wisdom from?

A Jansenist Friend

Arnauld was no stranger to mathematics. He had devised a new method for teaching children the principles of geometry. Yet he didn't solve mathematical problems and he didn't prove new theorems. A good friend of his, however, had devoted much of his short life to the art of mathematics. His name was Blaise Pascal.

After his vision of bliss, the former mathematical child prodigy had become a regular visitor to the monastery of Port-Royal. The two fellow Jansenists must have discussed all kinds of subjects, presumably also the subject of the uncertain future. Pascal had just solved the problem of the unfinished game by means of his *règle des partis* or, as we would say, the expected value. One can easily imagine Pascal explaining the concept of expected value to Antoine Arnauld in much the same way as it is explained in the last chapter of the *Port-Royal Logic*. What's more, at the conclusion of the chapter, we can almost hear Pascal speaking.

Pascal's Wager

The chapter ends with a thoroughly Jansenist message. One should always prefer a pious life to a worldly one, not on theological or philosophical grounds, as one would expect, but because of a mathematical argument.

In the eyes of the Jansenists the afterlife was not something you could earn. It was a grace from God. So from the perspective of us mortals it doesn't differ much from a future accident. Nor does the risk of being lost forever by leading a worldly life. The argument is a variation on Pascal's Wager that is discussed at length in his *Thoughts*. It goes as follows.

Pascal first states a principle that is known in mathematics as *the axiom of Archimedes*: it is in the nature of finite magnitudes that no matter how big they are, they can always be surpassed by smaller magnitudes if the number of these small magnitudes compensates for the difference in size.

This is the case when both something large and something small are weighed according to their respective probabilities and the greater probability of the small compensates for the difference in size. For instance, a 20% chance of winning 100 dollars outweighs a 1% chance of winning 1,500 dollars.

However, this does not apply to infinite magnitudes. These can never be balanced by finite magnitudes. So, no matter how small the chance of eternal salvation is, the expectation of the *infinite* afterlife always exceeds all the *finite* goods in the world taken together. Therefore, even if the chance of salvation is very small, one must still lead a pious life. And since all the temptations of this life do not outweigh the risk of being lost forever, however small this risk may be, one should never opt for a worldly life.

What Pascal means to say is that even if the chance of the infinite reward of an afterlife is very small the resulting expectation remains infinite and can never be balanced by the prospect of the many finite pleasures of a worldly life, since their magnitude will always remain finite.

So the *Port-Royal Logic* appropriately ends with a devout message that fits well with the central theme of the uncertain future: it is the greatest folly of all follies to spend your time and your life on worldly matters instead of what can help you to obtain the blessing of eternal salvation.

Pascal's Stamp

The spirit of mathematics that pervades the last chapter of the *Port-Royal Logic* and its call for the reader to lead a devout life unmistakably bear Pascal's stamp. One can almost hear him speaking, explaining the choices in life on the basis of the mathematical principle of the expected value. And to explain this principle he adopts the term *probability* that is discussed in the preceding chapters. In other words, Jacob Bernoulli may have introduced the term probability in mathematics, the creator of the mathematical term probability is most likely none other than Blaise Pascal!

The Golden Theorem

A Devout Family of Mathematicians

Jacob Bernoulli (1655–1705) was the first male descendant of what was to become the most famous dynasty of mathematicians in history. Until well into the eighteenth century the Bernoulli family would produce an unparalleled number of exceptionally gifted mathematicians. Three of them — Jacob himself, his brother Johann, and his nephew Daniel Bernoulli — are even among the greatest mathematicians of all time.

The Bernoulli family were Protestant refugees. They had fled Antwerp in the Low Countries and had settled in the Swiss city of Basel. There, Nicolaus Bernoulli earned his living as a merchant in spices and medical ingredients. He married the daughter of a Swiss banker, Margaretha Schönauer. Jacob was their eldest son.

At an early age Jacob goes to the University of Basel. Nothing indicates that he will become a famous mathematician. Quite the contrary. Under pressure from his parents, young Jacob devotes himself to the study of philosophy and theology. As a result, he learns to write Latin fluently and is also able to read ancient Greek. At just sixteen, he earns a degree in philosophy and, five years later, a master's in theology. But his real passion lies with mathematics and astronomy. Completely through self-study he manages to master both sciences.

Traveling through Europe

After finishing his studies, Jacob makes a journey through Europe. Every day he covers some forty kilometers. On horseback, on a mule or, if necessary, on foot. On his way he keeps a diary written in German. He calls it his *Reissbüchlein.*

His first stop is Geneva. To earn his living he becomes a private math tutor. One of his students is a very intelligent blind girl, Elisabeth Waldkirch. Although she is only fifteen years old, she already speaks fluent French, German, Italian, and Latin. Jacob successfully teaches her mathematics using a wooden board with letters and numbers carved into it. Later, this will inspire him to write an article about teaching mathematics to the blind. At the same time, he does not neglect his theological profession. Every day Jacob reads to Elisabeth from the works of a Swiss theologian.

After his stay in Geneva, Jacob travels on to Paris. There he stays with the followers of René Descartes. For no less than two years, he devotes himself to the study of the mathematical and philosophical works of the famous Frenchman. In his spare time, the young mathematician fanatically practices *jeu de paume*, a popular forerunner of modern tennis. He also writes poetry in several languages. On the occasion of his brother Johann's wedding he writes a long wedding poem in Latin.

The Comet of 1680

In the course of 1680 Jacob returns to Basel. At the end of the year, all over Europe an astounding phenomenon appears in the sky: an enormous comet moves along the firmament. The comet is so bright that it is visible even during the day. Perhaps most spectacular is its long tail, which can still be admired in many paintings from that time.

The theologians and many devout citizens of Basel interpret the impressive phenomenon as a sign from God. Not Jacob. He publishes an article in which he offers a purely physical explanation. According to Bernoulli, comets are nothing but celestial bodies moving in a fixed and therefore predictable orbit around some large distant planet. The comet of 1680, he predicts, will reappear in 1719.

It goes without saying that his theory is difficult to reconcile with the interpretation of the theologians. More or less as a concession, Jacob leaves it up to the theologians to determine whether or not the long tail of the comet is a sign of God's wrath.

Professor of Mathematics

Jacob won't stay in Basel long. After a year he packs his suitcases again and travels to the Low Countries, where he visits Leiden and Amsterdam. A year later he makes the crossing to England. In London he attends a session of the famous Royal Academy. He meets numerous leading mathematicians and scientists, like Robert Boyle and Robert Hooke, with whom he will continue to correspond.

At the end of the year he travels through Germany back to Basel, this time to settle there for good. Jacob becomes a teacher of experimental physics and marries Judith Stupanus, the daughter of a wealthy local pharmacist. They will have two children. A few years later he is offered an ecclesiastical appointment, which he renounces. Instead, at the age of 32, he accepts the chair of Professor of Mathematics at the University of Basel. He will hold the chair until his death.

Jacob is now an internationally renowned mathematician. He publishes numerous mathematical articles — the first scientific journals have by now seen the light of day — and his lectures in Basel attract students from all over the world.

A Fraternal Feud

One of Jacob's pupils is his twelve-year younger brother Johann. The tutoring is conducted in secret, because father Nicolaus wants his youngest son to become a merchant. Initially Jacob and Johann work brotherly together on the same problems. But when Johann steps out of the shadow of his older brother, the collaboration soon turns into a rivalry and the rivalry eventually boils up into a fraternal feud. They fight each other in articles they publish on the same problems. The brothers are both mathematical geniuses, but their characters are totally different. Johann is a quick thinker and impulsive. Jacob is contemplative and a perfectionist. The two will never reconcile.

The *Ars Conjectandi*

Once settled in Basel Jacob starts studying Huygens' treatise on the mathematics of games of chance. His earliest notes on probability can be found in his scientific diary and date from the winter of 1684–85. Eventually, his study will result in an extensive textbook on probability theory, *The Art of Conjecturing* (*Ars Conjectandi*).

JACOBI BERNOULLI,
Profeff. Bafil. & utriufque Societ. Reg. Scientiar.
Gall. & Pruff. Sodal.
MATHEMATICI CELEBERRIMI,
ARS CONJECTANDI,
OPUS POSTHUMUM.
Accedit
TRACTATUS
DE SERIEBUS INFINITIS,
Et EPISTOLA Gallicè fcripta
DE LUDO PILÆ
RETICULARIS.

BASILEÆ,
Impenfis THURNISIORUM, Fratrum.
clↃ Ↄcc xiii.

The first part contains Huygens' treatise and explains all of its problems. The second part contains Bernoulli's own mathematical theory of combinations, which he applies in the third part to in the analysis of various gambling games. One of these games is the infamous Bassette. This is a card game that has come over from Italy to France in 1674 and has become such a fad in higher circles that many an aristocrat has ruined himself playing it. It was banned by King Louis XIV who was deeply concerned about the survival of the French nobility.

Together, the three parts form an exemplary textbook on probability theory. Each problem is clearly formulated and each solution is proven in detail and provided with an appropriate example. Only the fourth and last part of the book still awaits completion.

Jacob works on the book in deep silence for twenty years. Few people know about the work, aside from some of his own students and his younger brother Johann. Then, in April 1703, a letter arrives from the great German mathematician and philosopher Gottfried Leibniz.

Jacob's Letter to Leibniz

The letter from Leibniz is mainly about differential and integral calculus. Jacob and his brother Johann had been the first to really understand and apply this initially rather obscure area of mathematics. The term *integral*, for example, was introduced by Jacob.

In a postscript Leibniz enquires about Jacob's calculations of *probabilitates*, "probabilities". Jacob is puzzled. How did Leibniz know about his project? He suspects that his brother Johann is the source. Leibniz denies this, but a letter from Johann to Leibniz from 1697 confirms that his suspicions were correct.

Jacob writes in return that he has indeed been working on a book about probabilities and that, in his opinion, there is hardly anyone who has devoted more thought to the subject. Nevertheless, his weak physical constitution makes writing quite laborious. If only he had a helper who could entrust all of his thoughts directly to paper! Although most of the book has been completed, he still has to solve a very difficult but extremely valuable problem. What was this problem?

Proving an Intuitive Law

"It is well known," Jacob explains, "that the probability of any event depends on the number of cases in which it can occur. That is how we know, for instance, that if we roll two dice, it is more probable that we will throw seven pips than eight pips. But we don't know, for instance, how much greater the probability is that a young man of twenty will survive an old man of sixty instead of the other way around. The reason for this is that we don't know in how many cases the young man will die before the old man, and vice versa."

"I began to wonder," Jacob continues, "whether we could get to know *a posteriori* ("in hindsight") probabilities that are unknown to us *a priori* ("in advance"). For example by observing a large number of pairs of young and old men."

"Suppose that in a thousand cases the young man survives the old man and in five hundred cases the old man survives the young man. Then one could conclude with sufficient certainty that the probability that the young man survives the old man is twice as high as the probability that the young man dies first."

"It is a miracle," notes Jacob, "that even the greatest idiot by some kind of natural intuition knows — by himself and without any prior training — that the more observations are made, the smaller the risk is that one will deviate from the goal (i.e. the probability one is looking for). But proving this in a strict and mathematical way is not at all easy!"

The Unfinished Masterwork

Jacob will never complete the ambitious fourth and final part of his mathematical masterpiece. Two years later, on August 16, 1705, after having struggled with poor health for some time, Jacob Bernoulli dies at the age of fifty, presumably from the complications of gout. Only many years later, in 1713, is the work finally published by Jacob's son Nicolaus (1687–1769), a talented painter whose portrait of Jacob can still be admired in Basel. The preface to the book is written by Jacob's nephew, also named Nicolaus Bernoulli (1687–1759), who himself has studied mathematics with his uncle.

The Law of Large Numbers

The law that, according to Bernoulli, is known even to the greatest fool is the famous *law of large numbers*. It was given this name in 1837 by the mathematician Siméon Poisson (1781–1840). The law can be easily illustrated by the tossing of a coin.

If you toss a coin only a few times, the result will often not reflect the true probabilities of heads and tails. For instance, if you toss a fair coin ten times, the outcome may very well be three heads and seven tails. In other words, only 30% of the time do you throw heads. However, if you keep tossing long enough, the average number of heads — often referred to as the *relative frequency* — will come closer and closer to 50%, its true probability.

Bernoulli was convinced that the law of large numbers, once proven, would also justify estimating probabilities that were not known in advance. For instance, the statistical probability of an old man dying earlier than a young man.

Bernoulli had worked secretly on a proof for over twenty years, and it was not in vain. The very last chapter of the *Ars Conjectandi* that Bernoulli managed to complete contains a strict, mathematical proof of the law of large numbers. Thanks to this proof, the *Ars Conjectandi* is recognized as a pioneering work in the history of mathematics. For the first time in history someone has succeeded in proving a general and fundamental law of probability theory. Before we outline Bernoulli's approach, we will briefly discuss what the law does and does not tell us.

A Sloppy Interpretation

Our intuitive notion of the law of large numbers is not always as clear and definite as one might hope. Ask people what will happen if they toss a fair coin a large number of times and the chances are that they will answer that the number of heads and the number of tails will increasingly balance each other out.

This is a rather sloppy interpretation of the law. In fact, the law of large numbers tells us that in the long run the *average number* of times you toss heads and the *average number* of times you toss tails will approach each other. In other words, the relative frequencies of heads and tails will get closer and closer to each other if you keep tossing the coin a large number of times.

The law does not say that the actual numbers of heads and tails will even out in the long run. Exactly the opposite is the case! If you toss a coin repeatedly and after each toss plot the actual number of heads or tails and the expected number of heads or tails (that is, half the number of tosses) in a graph, you'll see that ever larger and longer fluctuations will emerge.

It can be proven mathematically that the difference between the actual number of heads and the expected number of heads tends to grow proportionally to the square root of the number of tosses. In other words, the actual difference between the number of heads and the number of tails tends to grow. After all, the coin has no memory. Each toss is the beginning of a new series. There is therefore no law of compensation.

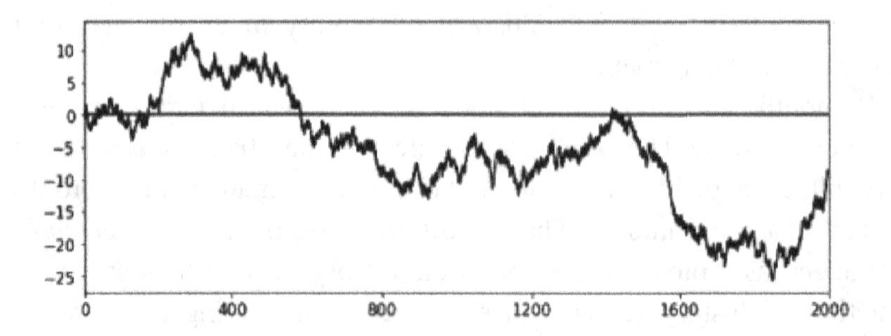

This counter-intuitive tendency is not in conflict with the law of large numbers. If you keep on tossing a coin indefinitely, the actual difference between the number of heads and the number of tails may very well tend to grow, while the difference divided by the number of tosses approaches zero. That is, the relative frequency of both heads and tails approaches 1/2.

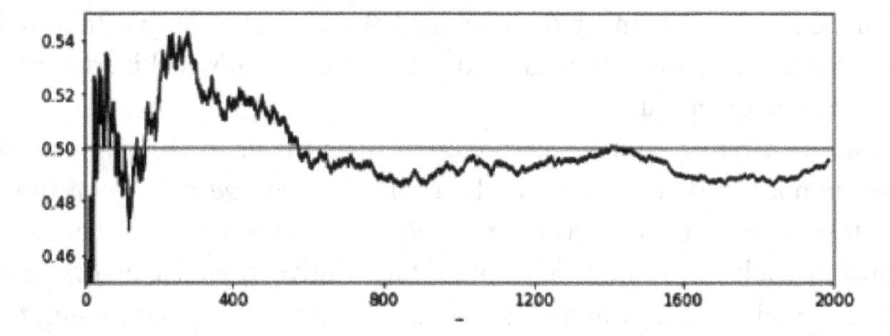

Bernoulli's Urn

How can something as intuitive as the law of large numbers be proven in a strictly mathematical way? Bernoulli did so by imagining a large urn filled with black and white pebbles that are well mixed. In our modern presentation we will use marbles instead, since marbles are indistinguishable from each other except for their color.

It is easy to see that the ratio between the number of white marbles and the number of black marbles in the urn determines the chance of drawing a black or a white marble from the urn. So Bernoulli's mathematical model enables us to create all kinds of different probability distributions, simply by varying the ratio between the black and the white marbles in the urn.

Imagine you are randomly drawing marbles from the urn one by one. Each time you jot down the color, put the marble back in the urn and mix it well with the other marbles. If you only draw *one* marble from the urn, there are of course only two possible outcomes. Either you draw a black one {B} or you draw a white one {W}.

If you draw *two* marbles consecutively, returning the first marble before you draw the second one, there are four (2 × 2) possible outcomes. Either you draw two black marbles {B, B} or two white marbles {W, W}, or you draw a black one and a white one {B, W}. The two marbles of the same color can of course only be drawn in one way, but the two marbles of different colors can be drawn in two ways: B W and W B.

If you draw and return *three* marbles successively, there are eight (2 × 2 × 2) possible outcomes. Either you draw three marbles of the same color, {B, B, B} or {W, W, W}, or you draw two marbles of the same color combined with one marble of the other color, {B, B, W} or {B, W, W}. The first can both be drawn in only one way, while the latter can both be drawn in three ways, as you can easily check yourself.

We can go on like this indefinitely, but by now you will probably have noticed that these numbers are exactly the numbers of Pascal's triangle!

Bernoulli Processes

In probability theory, this arrangement of possible outcomes is also known as the *binomial distribution*. In short, Bernoulli's mathematical urn is a universal model for all kinds of random processes that have the property that their possible outcomes follow a binomial distribution. In other words, they are arranged in the same way as the numbers in Pascal's triangle.

Examples are tossing a coin ten times to see how many tails occur; rolling a die to see if a 1 appears; buying lottery tickets, and so on. In general, these are random processes in which there are only two possible, mutually exclusive outcomes: success or failure, a yes or a no. They are called *Bernoulli processes*.

Bernoulli's Approach

Already in 1690 Bernoulli had formulated in his scientific diary exactly how he wanted to prove the law of the large numbers. Normally, it is not so easy to reproduce mathematical formulas and proofs in plain language, but fortunately the sketch in Jacob's diary is pretty straightforward.

Imagine there are 50 marbles inside the urn, say 30 white marbles and 20 black marbles. In this case, the probability of drawing a white marble is 30/50 (60%). Suppose you draw marbles at random from the urn, one by one, putting each marble back before drawing the next one. Take a small interval around the probability of drawing a white marble, for example, the interval between (30 – 1)/50 (58%) and (30 + 1)/50 (62%).

Now Jacob proved the following. The likelihood that the relative frequency of drawing a white marble — that is, the number of white marbles divided by the total number of marbles that you draw — falls within this interval can be made as high as one wants, as long as one keeps drawing marbles from the urn long enough. And this is true not only for our example, but for every ratio of white and black marbles and for any interval, no matter how small the interval.

For instance, if you toss a fair coin, Bernoulli's mathematical version of the law of the large numbers says that the average number of heads (or tails) comes as close to 50% as you want, with a likelihood as high as you want if you keep tossing long enough. This corresponds exactly to our intuitive version of the law that says that if you continue to flip a coin, the average number of heads (or tails) gets closer and closer to 50%.

A Golden Theorem

The importance of the mathematical law of large numbers can, indeed, hardly be overestimated. In its intuitive form, the law of large numbers is fundamental to our understanding of the concept of probability. In

Bernoulli's words, even the biggest idiot knows the law. Its mathematical formulation provides a link between mathematical theory and practice.

According to Abraham de Moivre (1667–1754), one of the most important mathematicians in the history of probability theory, Bernoulli's proof is of great beauty. Bernoulli himself considered his proof for the law of large numbers to be even more important than a proof of the quadrature of the circle. He called it his Golden Theorem.

How Large is Large?

Bernoulli's law states that you get a good approximation of the probability of drawing a white marble if you keep drawing *long enough*. But what does Bernoulli mean by "long enough"? How long do you have to flip a coin to get close to a fifty-fifty ratio? Here, our intuition leaves us completely in the dark. We just have no idea how large the "large" of the law of large numbers is. But Bernoulli's proof also enabled him to find an answer.

Let's take the same example as the one Bernoulli himself used. Again, the urn is filled with 50 marbles, 30 white ones and 20 black ones. So the probability of drawing a white marble is still 30/50 (60%). Let's take the same small interval as we did above, the interval between (30 − 1)/50 (58%) and (30 + 1)/50 (62%). Again, we draw marbles at random from the urn, following the same procedure as above.

Say we want it to be a thousand times more likely that the relative frequency of drawing a white marble falls within the chosen interval than outside it. In other words, we want it to be 99.9% (1000/1001) certain that the average number of times that we draw a white marble falls between 58% and 62%. How many times do we have to draw a marble from the urn?

Bernoulli calculated that, at the utmost, we have to draw a marble 25,500 times! This is probably far more times than you ever imagined. It is, as the statistician Stephen Stigler remarked, more than the number of residents of Basel in those days.

The upper limit of 25,500 calculated by Bernoulli is, we have to admit, on the rough side. Inspired by Bernoulli's proof, Abraham de Moivre derived another basic theorem of probability theory — today known as the *central limit theorem* — which enabled him to calculate the value of the upper limit exactly at 6,498. A lot less than the figure arrived at by Bernoulli, but it would still take you a considerable amount of effort and time.

Flipping a Coin

So the "large" in the law of large numbers is much larger than we intuitively imagine it to be. Just flip a coin for a while and you'll see that the averages of both outcomes fluctuate considerably. Only in the very long run will the average numbers stabilize around their probabilities, as can be illustrated by some mathematics.

Suppose you want the average number of times you toss heads to be somewhere between 48% and 52%. If you toss a coin *ten* times, it can be calculated that the likelihood of this is only 25%. This probably won't amaze you. It is in line with our experience.

But what if you toss the coin a *hundred* times? Sure, you might think, the likelihood will increase significantly. Not so. With a hundred tosses, it's still only 38%. In other words, it is still more likely that the relative frequency of heads falls outside the interval we've chosen than inside it. And this you probably wouldn't have expected.

How about tossing the coin a *thousand* times? That raises the likelihood considerably. But even then there still is a chance of about one in five that the relative frequency of heads falls outside the chosen interval than inside it. That is 20% of the time.

As these numbers show, it will take you a lot of time and dedication to calculate probabilities using the law of large numbers, as Bernoulli had in mind. However, nowadays we have a device that has no problem with these large numbers: the modern computer.

Computer Simulation

No human can measure up to the speed at which a computer can perform a large number of calculations. If we could simulate chance events such as the rolling of a die on the computer, we would have an adequate estimation of the probability of any possible outcome in no time at all. But how do you program a computer to simulate the ways of chance?

Somehow the computer has to generate random sequences of numbers, like the outcomes of a die. If we can come up with a formula that produces a sequence of numbers that is statistically indistinguishable from a random sequence, then we could program the computer to be its own random number generator. What we need is a *recursive formula.*

A recursive formula is a formula that uses its output as input to calculate the next number. It works as follows. First, you choose a starting number, known as the *seed.* Then you apply the formula to this number. The outcome is the first number in the sequence. This first number is then used as new input for the formula to generate the next number in the sequence, and so on.

The only problem is that as soon as a certain number reappears, the same sequence of numbers repeats itself all over again. The *Mersenne twister generator*, which was developed by Japanese mathematicians at the end of the last century, has such a long cycle that when it starts repeating itself mankind will be long gone.

So all you have to do is to find a way to simulate a specific chance event, such as the rolling of a die, and let thc computer do the work. This is quite simple. You first have the computer generate a random number between zero and one (the standard interval of the random number generator). Then, you instruct the computer to multiply this number by six and round up the resulting number to an integer between one and six. Finally, you enter how many times you want to repeat the simulation.

You can simulate any random process you want. For example, you can use the computer to simulate the problem of the unfinished game with multiple players. Or if the odds of winning are not equally divided. In this way many problems that require complex calculations can be solved in no time at all.

With a giant leap we have entered the modern era of the computer and come to the end of our journey through the early history of probability theory. The concepts and rules of probability theory that we have introduced so far are relatively easy to grasp. The developments in probability theory that have taken place since the time of Jacob Bernoulli require some serious knowledge of mathematics and therefore fall outside the scope of this book. Fortunately, you now possess all the basic knowledge you need to unravel the counter-intuitive logic of chance in Part Two.

Part Two

❖❖❖

The Logic of Chance

1 The Capricious Ways of Chance

The Gambler's Fallacy

Modern Gambling

The days are long gone when people met in dark pubs or fancy salons to spend their evenings playing dice or cards for money. Gambling no longer appears to be as popular a pastime as it once was. At the same time, however, the amount of money that changes hands in the world of modern gambling tells a different story. Gambling has become a billion-dollar industry. It is an industry that thrives on large sums of money and, as we will see, on even larger illusions. At the heart of this industry is the modern-day palace of gambling, the casino.

The Casino

The oldest known casino in the world, the *Casino di Venezia*, opened its doors in 1638. Here, Casanova experienced many of his amorous adventures. But it wasn't until the nineteenth century in Europe that the casino as we know it today emerged. Napoleon, a notorious player himself, was the first to legalize gambling. Initially, gambling was only allowed in a number of rooms of the *Palais-Royal* in Paris. However, it didn't take long before some real gambling palaces arose.

The first casinos, as they were called, not only facilitated gambling, but also offered luxury dinners and all sorts of entertainment to the rich of Europe. In 1841, the French mathematician and banker François Blanc and his twin brother Louis opened a grand casino in the German spa town of Bad Homburg. In no time the spa became extremely popular with the Russian nobility.

It inspired Princess Caroline of Monaco also to open a casino in Monaco. The gambling palace was meant to rescue the Grimaldi family, which ruled over the small state, from imminent bankruptcy. After gambling was banned in Germany around 1860, François Blanc was put in charge of the Grimaldi project.

The banker managed to attract some important investors, including the bishop of Monaco and also the later Pope Leo XIII. Once he had raised enough money, François and his brother Louis proceeded to transform Monte Carlo from an insignificant capital of a small principality into a metropolis of luxury, entertainment, and gambling. In the 1870s, the *Grand Casino de Monte Carlo* was already attracting no fewer than 120,000 visitors a year. Nowadays, the casino is still the Grimaldi family's primary source of income.

The luxurious gambling palace in Monaco was the example *par excellence* for the American casinos. In the 1940s the first gambling palaces opened their doors in Las Vegas. They were run by the mafia for quite some time and Las Vegas soon became known as the *City of Sin*. Once nothing more than a desert, the Las Vegas Valley grew into a veritable gambling Valhalla, with no fewer than 122 casinos, each of them offering its visitors all kinds of entertainment in addition to gambling.

Today, you'll find casinos in almost every part of the world and there is a multitude of casinos active online. The casino's success is shown by the amount of money that changes hands in a year. In 2015, casino visitors in the United States alone spent a total of 357 billion dollars. This financial success story is, as we will see, mainly due to a deeply rooted mathematical delusion, known as the *gambler's fallacy*.

Roulette

In American casinos craps and blackjack are most popular, but in casinos outside America roulette was and still is the queen of the games of chance. Roulette means "small wheel" in French. Tradition has it that none other than Blaise Pascal invented the predecessor of this little wheel while attempting to construct a perpetual motion machine. But it would not be until the rise of the casino in the nineteenth century that the game of roulette as we know it today would begin its triumphal march. Once it had conquered Europe, it crossed the ocean to America.

The numbers 0 to 36 are located on the rotating wheel, each with its own pocket. The pockets are distributed in a circular way over the wheel. Eighteen of the numbers are red, and the other eighteen are black. The zero is green (the American roulette wheel differs slightly, for it also has a double zero).

Players bet on the outcome of the spinning wheel. They do so by placing chips that represent a certain amount of money on the game board. On this board the numbers 0 to 36 are arranged in such a way that one can bet not only on each individual number but also on all kinds of combinations of numbers. In addition, one can bet on *red* or *black* and on *even* or *odd*.

The *croupier* conducts the game and, most importantly, operates the wheel. *Faites les jeux*: the players are allowed to place their bets. *Les jeux sont faits*: the croupier spins the wheel and flicks the ball in the opposite direction along the inside edge of the wheel. *Rien ne va plus*: no more bets are allowed. The roulette ball spins, dances, and jumps over the rotating wheel until it finally comes to rest in one of the pockets. The corresponding number is the winning number.

You don't have to be an experienced player to play roulette. Its rules are quite simple and its outcomes are completely random. Chance rules, and every player is completely at its mercy. Nonetheless, many have tried to understand the capricious ways of chance and have come up with some kind of system that supposedly beats chance. In vain. The magic of the spinning wheel, in combination with the many betting possibilities and the simplicity of the rules, has given roulette its extraordinary appeal — an appeal that, as we will see, can turn into a fatal attraction.

Monte Carlo, August 18, 1913

In the summer of 1913 at the *Grand Casino* in Monte Carlo, the worldly elite of old Europe, in impeccable evening dress and with an aristocratic smile, meets in the gaming rooms. For these wealthy visitors gambling is an entertaining way to pass their evenings. The visitors of lower standing, on the other hand, play to get rich quick or to pay off a heavy debt. Their hopes can be read on their faces. But on the evening of August 18, everything is different.

In the roulette room the air is filled with excitement. The public has crowded around the roulette table, anxious to catch a glimpse of the extraordinary course of events. Sixteen times in a row the ball has come to rest on a black number! Those who are lucky enough — or, as it turns out, not so lucky — to be close to the table are obsessively and incessantly betting their money on *red*. A common conviction has taken hold of the visitors: it can't be long before a red number will show up.

Aristocratic self-control has given way to a feverish look in the eyes and ever-larger sums of money (there were no chips yet) are placed on *red*. From the back of the crowd, people who can't see anything or are unable to place their bets get noisier and noisier and try to find their way to the roulette table. Time and space, the whole universe seems to be centered around the roulette table.

Barely audible, the *rien ne va plus* of the croupier sounds. The din subsides. Everyone is in anticipation of the outcome of the spinning wheel. *Noir*. Black again! Smiles fade and faces turn pale in the front row. Disbelief surges through the room. Some whisper that the devil himself must be spinning the wheel!

So it goes the eighteenth time, and the nineteenth time, and so on. Each time the ball comes to rest on a black number. People place towering stacks of gold pieces on *red* to make up for their now excessive losses. With every spin of the wheel they become more and more convinced that a red number must and will fall. Yet the piles of gold on the table vanish as quickly as they grow. Eventually, after twenty-six times in a row, the memorable series of black outcomes comes to an end.

On that evening entire fortunes were lost. Meanwhile, the Grimaldi family had become millions of French francs richer. All this happened because most visitors were convinced that after a long series of black numbers a red number was almost guaranteed to fall. It was a remarkable evening in the history of gambling, but not unique. In fact, not so long ago, another game of chance drove almost an entire country to a similar form of collective madness.

Lotteries

Although a lottery may not be a game of chance in the common sense of the word, it is certainly a form of gambling ruled by chance. In a lottery a large number of participants purchase lottery tickets for a small amount of money each. Each lottery ticket has a specific combination of numbers. The winning combinations are selected by a random draw of numbers. Typically you buy an extremely small chance of winning an extremely large amount of money, called *the jackpot*.

Lotteries were and still are used to raise money for government projects or charities. The oldest known lottery was held in China at the beginning of the Han dynasty (about 200 BCE). In the West, the Roman emperor Augustus was the first to use a lottery. According to the ancient historian Suetonius, Augustus sold lottery tickets at dinner parties to his guests, presumably to pay for repairs in the city of Rome.

The first documented lotteries with money prizes date from the fifteenth century. Cities in the Low Countries used lotteries to raise money for fortifications and for the poor, cities in Italy to wage war against their rivals. In the seventeenth century lotteries were a common means to collect money to fund public projects, especially in the Netherlands. The English word *lottery* goes back to the Dutch "lot".

The Dutch State Lottery is still running, but most lotteries did not survive the nineteenth century. Lotteries were considered a hotbed of financial and moral corruption. The English State Lottery was abolished in 1826, and by 1890 lotteries were prohibited in the US in every state except Delaware and Louisiana.

The twentieth century, however, saw a worldwide rebirth of lotteries. France reintroduced its national lottery in 1933, while the first American lottery was re-established in Puerto Rico in 1934. Many states legalized state lotteries in the 1980s, the most recent being Mississippi in 2018. In the United Kingdom the first draw of the National Lottery took place on 19 November 1994. It was broadcast live on television.

Today, lotteries are the most widespread form of gambling. In 2018 the state lottery sales in the United States alone totaled 85.59 billion dollars. What is it about lotteries that makes them so popular? Commercials selling false hopes and the live broadcasting of the draws on television have firmly anchored lotteries in modern popular culture. But people's main reason for participating in lotteries has always been the dream of winning the jackpot.

For a small amount one buys a chance of winning a huge amount of money. Your chance of becoming a millionaire! Top prizes of tens of millions are no exception. The largest jackpot ever was 1.586 billion dollars. On January 13, 2016 this incredible amount was paid out in the US lotto game Powerball.

A huge jackpot sells, however small the odds of winning the jackpot may be. As Arnauld, or rather Pascal, already observed. For instance, the chance of winning the Powerball jackpot is about 1 in 146 million. To give you an idea, you have a better chance of throwing heads twenty-seven times in a row.

A Collective Psychosis

In Italy, lotteries are a national pastime. Every Wednesday and Saturday evening millions of Italian families gather around their televisions to watch the Italian State Lottery. The lottery is held in ten different major cities across the country. In each of these cities five numbers from 1 to 90 are drawn. People can bet on one, two, three, four, or five numbers. There is no limit to the amount of money that can be bet.

The Wednesday evening of February 9, 2005 is etched on the memories of many Italians. That evening, in the lottery at Venice, after no fewer than 182 draws, the number 53 was finally drawn once again. In the preceding year and a half, almost all of Italy had lived under the spell of this magical number. For months it had eluded every draw. With each new drawing of the Venice lottery, more and more people were convinced that this time the number 53 was bound to be drawn. How could it not be?

Mathematicians appeared on television, warning the public that their belief was based on an illusion. To no avail. People were fully convinced that the number 53 ultimately had to appear. Self-proclaimed experts and clairvoyants were consulted en masse. Still many people incurred heavy debts and quite a large number of them went bankrupt and lost their homes.

Even some suicides were committed. A woman from Tuscany drowned herself in the Tyrrhenian Sea after having spent all of her family's money. A man from a village near Florence first shot his wife and son, then himself. In a desperate response, a national consumer rights association called for the number 53 to be banned from the lottery. This collective psychosis had to stop!

When on the memorable evening of February 9, 2005, the number 53 reappeared for the first time in 182 draws, the Italian state had to pay out some six hundred million euros to an unknown number of winners. Yet it is estimated that in the previous months Italians, by betting ever larger amounts of money on number 53, had lost more than three and a half billion euros to the State.

The Gambler's Fallacy

The delusion that captivated the minds of huge numbers of Italians, as it did those of the distinguished public at the Monte Carlo casino, is known in mathematics as the *gambler's fallacy*. A fallacy is a way of a reasoning that *seems* logical but is not logical at all. It is a deceptive kind of reasoning that sometimes can almost take the form of a delusion, as is the case with the gambler's fallacy. Even if its reasoning is shown to be invalid, the gambler's fallacy still keeps exerting a strong influence on our minds.

The fallacy goes like this: when a chance event occurs less often than average during a certain period of time, the probability inevitably increases that this event will occur in the near future, even though the events are independent. This fallacious reasoning is especially common among gamblers. Hence its name.

For example, when you flip a coin and throw a sequence of tails, you probably will expect that heads is bound to appear next. Another well-known example, from outside the context of gambling, is the popular belief that after one has given birth to three daughters, the fourth child is bound to be a son. Everyone will recognize the almost ineluctable spell of this fallacy.

The Absurdity of the Fallacy

A thought experiment will give you an idea of how absurd this delusive fallacy is. Imagine you happen to walk into a room when someone is tossing a coin. He or she asks you to guess heads or tails. Heads, you say. You could as well have chosen tails, for the odds are fifty-fifty. But then the other says "Are you sure? I've just tossed heads six times in a row!" Does this really all of a sudden change the odds? No, of course not!

When outcomes are independent of each other, as in the examples mentioned above, previous outcomes have no influence whatsoever on the probability of a subsequent outcome. Neither the roulette ball nor a coin nor a die has any memory of previous outcomes. Each spin of a roulette ball, each toss of a coin and each throw of a die is the beginning of a *new* series of outcomes. A series as erratic and unpredictable as any other series of outcomes. The odds are and remain exactly the same.

Take, for instance, the roulette. There are still eighteen red and eighteen black numbers on the roulette wheel, and the ball spins no less unpredictably over the rotating wheel than it did in the previous rounds. So the chance that at the next spin of the wheel the dancing and jumping ball will come to rest on a black number remains 18/37. And the same holds true for the chance that it will come to rest on a red number.

Many visitors to the casino are convinced that once they've lost money at the roulette table, they can make up for their loss as long as they keep playing until "the tide turns". It's an illusion. In the end, the casino is the only one that benefits. The gambler's fallacy keeps visitors at the roulette table, and because the casino has a house edge of 2.7% for every possible bet, the more and the longer the visitors gamble, the more profit the casino will pocket. The law of large numbers is the engine, the gambler's fallacy its fuel.

Mathematical Roots

The gambler's fallacy has a captivating logic of its own, one that is hard to escape. Even if we know better, we still tend to fall for its deceptive reasoning. How can a fallacious intuition wield such an unprecedented power over our minds? Undoubtedly, psychological factors play a role, but in essence the fallacy is a mathematical one.

We all have an intuitive understanding of the law of large numbers. As Bernoulli wrote, even fools with no education whatsoever. At the same time we have no idea how large the "large" of the law really is. In reality, the law only applies in the long run, way beyond what we are used to imagining. The numbers of outcomes that we have in mind are far too limited. And herein lies the root of the gambler's fallacy.

Intuitively, we imagine the law already applies to numbers that are still within our grasp. In other words, we believe that the underlying probabilities are already reflected in the average of the outcomes that we observe in the short run. This is why we are convinced that an "excess" of black outcomes must be compensated for by red ones and vice versa.

If black numbers are not in a more or less proportional way alternated with red numbers, according to our naïve understanding of the law of large numbers, the roulette wheel would have a preference for black numbers. But this cannot be. After all, there are as many black numbers as there are red ones.

We conclude that in order to compensate for this "excess" of black numbers, a red number is bound to follow. In this way we arrive at the fallacious belief that the longer a series of black outcomes is, the more probable it is that a red outcome will be next. This "chain of logic" makes the gambler's fallacy such a compelling thought. And yet no such tendency or law of compensation exists. As said before, the roulette ball has no memory of previous outcomes.

The reason that the averages of the outcomes ultimately move towards the underlying probabilities is not because runs of the same outcome are compensated for. It is because *in the long run* runs of the same outcome are, as it were, absorbed into the average of all the outcomes.

Take, for example, the throwing of heads and tails. As we have already seen, the *actual* difference between the total number of heads and tails, as opposed to the *average* difference, tends to show bigger and bigger fluctuations. So instead of a tendency to compensate, the difference between the outcomes tends to increase.

The prominent psychologists Amos Tversky (1937–96) and Daniel Kahneman (1934–) named the persistent misconception that probabilities are reflected in the averages of limited numbers of outcomes *the belief in the law of small numbers*. They considered it to be the result of a specific cognitive-psychological mechanism.

As we have observed, the belief is rooted in a lack of mathematical intuition. We simply can't wrap our minds around how large the numbers in the law of large numbers actually are. Paradoxically, a basic law of probability is the reason why an fallacious belief exerts such an ineluctable power over us.

Streaks and Runs

Real or Faked?

Suppose a fair coin has been tossed a hundred times. Below are two sequences of one hundred consecutive outcomes. One of these two sequences is the true one, while the other is an attempt by someone to imitate a random series. Can you tell which of the two sequences is the real one?

THTHTTTTTHTHTTHHTTTH	TTHTTTHHTHTHTTTHHTHT
HHTHHHTHTTTTTTHTHTHT	THHTHTHTHTTTHTHTTHTH
TTHTTHTTHHT HHHTHTHTT	HHTTHTHTTTHHTTHHHTTH
THTHHTTTTHTHHHHTHHTH	HHTTHHTTHHHTHTHHTHTT
HTHTHTHTTTHTTTTTTHTT	TTHTHHTHTHTHHTTTHHHT

Intuitively, you may be inclined to consider the one shown on the right to be the true random series. Why? In the sequence shown on the right heads and tails alternate in a fairly balanced way. In the sequence shown on the left, there are quite some long runs of the same outcome, far more than in the sequence shown on the right.

But if you therefore consider the sequence shown on the right to be more representative of a random series, then, once again, your intuition is fooling you. We tend to imagine random series to be much more "beautiful" than they actually are. For exactly the same reason as we are prone to commit the gambler's fallacy. We expect that a sequence of a hundred tosses will reflect the underlying probabilities.

The sequence shown on the right has 48 tails and 52 heads. In fact, as we previously observed, with a hundred tosses the likelihood that the average number of times you toss heads will be somewhere between 48 and 52 is only 38%. As a rule, a sequence that is truly random will have much less alternation between its outcomes and be a lot less balanced. In other words, there would be more and longer runs of the same outcome. So the real random series is the sequence shown on the left.

The Professor's Trick

A professor of statistics and probability theory once asked his students, for homework, to flip a coin a hundred times and to write down the outcomes, heads or tails. Some of the students dutifully carried out the experiment as requested. But a number of them thought they could save time and fool the professor by simply inventing a "random" sequence of heads or tails. The next day, the professor only had a quick look at the results. To everyone's surprise, he immediately identified most of the faked ones. Now how did the professor know?

We humans do not possess an internal random number generator that enables us to write down a random series of outcomes. Instead, we look at what we have written down and then determine how to continue in order to obtain a sequence that looks random to us. Long runs of the same outcome do not seem likely to us, so we alternate the previous outcomes.

But since a coin, however, has no memory of previous outcomes, it will exhibit no such balancing behavior. As a result, we greatly underestimate the number of runs of one and the same outcome. At the roulette table, for instance, we believe that runs of only red or black numbers are the exception. "*I've lost five times in a row on red. Too bad!*" A frequent alternation of outcomes appears "normal" to us. But contrary to what we think, runs of one and the same outcome are quite characteristic of random sequences.

What's more, we not only underestimate the number of runs of one and the same outcome, but also the length of these runs. Intuitively we avoid any runs longer than four. The reason is simple. Longer runs seem to contradict the law of large numbers. They give us the impression that the coin has a preference for one of the outcomes.

It is true that the probability of getting five heads or tails in a row in five tosses, as you can easily calculate for yourself, is only 1/16, or a little over 6%. But in one hundred coin tosses, a run of five of the same outcome can start at 96 different points. In other words, there are 96 opportunities for a run of a length of five to occur, and this explains why a run of five is not that unlikely at all in a sequence of one hundred tosses.

The probability that runs of a certain length will occur in a specified number of coin tosses can be calculated using an easy logarithmic formula. If you flip a coin a hundred times, the probability of getting the same outcome at least five times in a row turns out to be no less than 97.2%! In other words, the probability that no run of five or more will occur is a meager 2.8%. Contrary to what we think, it would be an exceptional event if such a run did not occur.

Even longer runs are quite likely to occur in a sequence of one hundred tosses. The chance of a run of six or more occurring is 80.7%, and the chance of a run of seven or more is still 54.2%. And these are only the runs that are more likely to occur than not, but even a run of length eight or more still has a 31.5% chance of occurring.

These findings will probably surprise you. They confirm that runs of the same outcome are, in fact, characteristic of random sequences and, moreover, that these runs tend to be much longer than we intuitively imagine them to be. They are a consequence of the unpredictability and independence of chance events.

Now we can understand how the professor was able to spot the majority of the faked random sequences. He only had to check the students' results for the presence of long runs of the same outcome. When you flip a coin a hundred times, the longest run of the same outcome you may expect to find is a run of length six or seven. Yet most of us avoid any runs longer than four. And so did the students who thought they could fool the professor.

The Hot Hand

Not only do we not associate long runs of a same outcome with chance, our brains are also programmed to see meaningful patterns everywhere. This fatal combination often blinds us to the role chance plays in the world around us. A classic example is the belief in the *hot hand*.

Imagine you are one of the star players in the NBA, the world's premiere basketball league. You are playing a decisive match. The atmosphere in the arena is electric. You have just scored four times in a row. The spectators are chanting your name louder and louder, and yet you are in a serene state of mind as if you were in a higher realm. With every basket your self-confidence increases. It is like you have wings.

You have a clear view at the basket and you launch the ball in one smooth movement. The three-point attempt hangs in the air. The spectators, players on both the court and the bench, and the coaches all anxiously follow the arc of the ball. Not you. Even before the ball has left your hands, you know it will go in. You've got a hot hand tonight!

The *hot hand* is a phenomenon that has been inextricably linked to basketball since its early days. The shots of an offensive player often occur in *streaks*, that is, continuous series of either hits or misses. If a player has scored two or three times in a row and then makes another basket, he is said to have a *hot hand*. If, on the other hand, a player misses again and again, he is said to be *cold*.

The hot hand is a firm part of the psychology of basketball. In sports commentary, you will hear about the hot hand time and again. "Player A's hot hand led his team to victory". If a player has a hot hand, many believe there is a good chance he will also score on his next attempt.

Random Streaks

In the mid-1980s an article was published that led to a debate that continues to this very day. The title of the article was *The Hot Hand in Basketball: On the Misperceptions of Random Sequences*. Researchers Amos Tversky, Thomas Gilovich, and Robert Vallone wondered whether the belief in a hot hand can be justified. And, in particular, whether there is truth in the belief that someone's chance of scoring increases as a result of their having a hot hand. Hundreds of basketball fans were interviewed and an overwhelming majority (91%) were convinced that it did. The researchers were not.

Imagine a professional basketball player with a scoring rate of 0.5 (50%). An offensive player has as a rule some twenty shots per game. Suppose for a certain game the shots of our player are represented by the following sequence:

X O X O X X X X X O X O X X O O X X X O

An "X" is a made basket and an "O" is a miss. Does this player have a hot hand?

He makes five baskets in a row once and three baskets in a row once. He is successful in thirteen out of twenty attempts and he even makes seven out of his first ten attempts. Only once does he miss twice in a row. It certainly looks like this player has a hot hand. But if you take a closer look at the sequence, you may notice that it is the same as the first twenty outcomes of the coin tosses we discussed above! A hit corresponds to tails, a miss to heads.

The researchers observed that the number of streaks of four, five, or six hits made by players with a scoring rate of 0.5 was no greater than one expects in a random sequence of heads or tails. Suppose you flip a coin twenty times. Your chance of getting heads three times in a row is 78.7% and your chance of getting heads four times in a row is 47.8%. Even your chances of getting heads five or six times in a row are still 25.0% and 12.2%, respectively. In other words, there is a big chance you will have a "hot hand" when you flip a coin!

The Independence of Hits

The researchers also observed that players had no greater chance of scoring after one, two, or three successful shots than if they had just missed one, two, or three times in a row. In addition, no player was able to tell in advance, depending on whether he thought he had a hot or a cold hand, how likely it was that the next shot would be a hit or a miss.

It is important to realize that this does not mean that the performance of players is purely random, as the researchers rightly point out. It depends on many non-random factors, such as the abilities of the offenders and the defenders, strategy and tactics, and so on. Top players will have longer and more streaks than mediocre players, though of course this is in line with their scoring percentage.

The crucial observation is that a hit or a miss does not depend on any previous attempts. Just as in a random sequence, an outcome is independent of, and cannot be predicted from, previous outcomes. In short, there is no statistical evidence for the existence of the hot hand. The opposite is true. The sequence of hits and misses does not differ from a random sequence. This is only logical, since the court situation is different in the case of each scoring attempt.

An Unyielding Belief

Why, then, do many people insist on the existence of a hot hand? One can think of at least two reasons. One is that our brains are programmed to find meaningful patterns in everything. Many people adhere to a form of popular psychology that says that with every hit a player's self-confidence and ease of play increase. The player *gets into the zone*.

Another explanation is that people erroneously believe that relatively long streaks are not characteristic of random sequences, as we observed above. Just like the gambler's fallacy, the belief in the hot hand is a consequence of our intuitive misunderstanding of the "large" in the law of large numbers.

The debate on the existence of the hot hand continues. In the world of basketball the hot hand is still a widely accepted phenomenon. When the former coach of the Boston Celtics was confronted with the above research, he commented: "Who is this guy? So, he makes a study. I couldn't care less".

Does not believing in the hot hand really diminish your enjoyment of the game? It would detract from the heroics of basketball, but the series of consecutive hits and misses would be no less exciting.

Chance Fluctuations

The Movement for Smaller Schools

Another consequence of our applying the law of large numbers on sequences or samples that are far too small ("our belief in the law of small numbers") is our insensitivity to the random fluctuations that occur in relatively small samples. In our search for meaningful patterns we all too quickly mistake these fluctuations for important psychological or sociological phenomena.

In the 1990s, Bill and Melinda Gates launched a noble initiative. They freed up huge sums of money from their Bill and Melinda Gates Foundation to encourage splitting large schools into smaller schools across the United States. Over the previous few decades Americans had become more and more dissatisfied with the ever-increasing size of schools. There was a growing belief that the smaller schools of the past had been better equipped to offer high-quality education and that, consequently, pupils would perform better if they went to smaller schools.

The statistics seemed to endorse this. Small schools were clearly over-represented among the best-performing schools. Almost one-third of the schools in the smallest 10% of schools were in the top 25. At the other end of the spectrum, only 1.2% of the largest 10% of schools were represented in the top 25.

At first glance, Bill and Melinda Gates' initiative appears to have been a praiseworthy enterprise. Numerous foundations and funds followed suit, including the Department of Education. In 2001, Bill and Melinda's foundation had already invested the enormous sum of 1.7 billion dollars, leading to the opening of 2,600 new small schools in 45 different states.

Unfortunately, contrary to expectations, there was no measurable increase in the performance of the students. In fact, students and parents began to complain about the limited selection of courses offered and the small number of teachers at these small schools. In their optimism, the educational reformers had overlooked something quite essential. They had only paid attention to the top 25 of the best performing schools. But smaller schools were also over-represented among the worst performing schools! Why?

Random fluctuations

One will look in vain for a psychological or sociological explanation of this surprising phenomenon. The fact that some of the smaller schools scored remarkably well and others performed very poorly is largely due to chance. The smaller the numbers, the larger the fluctuations. The capriciousness of chance accounts for small schools ending up among both the best and worst performing schools. A simple example will show why this is the case.

Suppose you roll a die, not once but several times. Now, observe what happens to the average score of the throws. Obviously, your average score will be somewhere between one, the lowest possible score, and six, the highest possible score. It will be one if you throw only ones. It will be six if you throw only sixes. All other possible outcomes will result in an average score that is in between these numbers.

Now, you are far more likely to get a high or a low average score when you only throw twice than when you throw ten times in a row. It is easy to see why. The chance of getting either two ones or two sixes in only two rolls is a lot greater than the chance of getting either ten ones or ten sixes in ten rolls. As a consequence, one can expect bigger fluctuations in the average score between samples consisting of a few rolls than between samples that consist of many rolls.

Likewise, small schools only need a few excellent students to enroll for their average score to skyrocket, while just a few poor students (or poor teachers) will cause their average scores to drop sharply. The fluctuations of the average score will be greater among small schools than among large schools. Just as we observed in the above example of throwing a die.

Intuitively we apply the law of large numbers. Therefore we believe that the averages will already stabilize for limited numbers of outcomes. Accordingly, we start looking for a meaningful explanation of results that are only due to chance fluctuations.

If the performance of these same schools is reassessed a few years later, the same picture as before is likely to emerge. Except that this time presumably some other small schools will show up among the best performing and some other among the worst performing schools. All due to chance fluctuations.

Misleading Mortality Rates and Unreliable Crime Rates

A similar blindness to random fluctuations often shows up in monitoring the quality of hospitals and in fighting crime in cities. If one looks at the mortality rates related to operations, small hospitals are more likely than large hospitals to be either at the top or the bottom of the rankings. Again, this is not necessarily the case because the surgeons in smaller hospitals are better or worse than their colleagues in larger hospitals. Small hospitals are just far more susceptible to random fluctuations in the number of mortalities than large hospitals are. Exactly as we observed in the case of small schools.

The same holds true for the crime rates of cities. A mayor of a small town would do well not to base his enforcement policy on the crime rate alone, as the number of murders, for example, could vary greatly in terms of percentage from year to year. A large percentage increase in murders would not necessarily mean that the town has become less safe. Most of the time, what we are actually measuring is fluctuations that are due to chance alone.

Sense and Nonsense of Psychological Research

It is to Bill Gates' credit that he took the blame himself and became more realistic in his pursuit of better education. Bill and Melinda's foundation has now wisely shifted its attention to what is happening in the classroom itself.

In the research commissioned by the Gates, the researchers had only observed that smaller schools were relatively well represented among the best performing schools. The fact that they were also well represented among the worst performing schools was apparently not taken into account.

What were actually fluctuations due to chance became interpreted as a socio-psychological phenomenon. The psychologist Kahneman refers to the tendency to underestimate the random variation in limited numbers of outcomes as an "insensitivity to sample size". It has the same roots as the "belief in the law of small numbers".

Especially in psychology, this insensitivity is rather widespread among students and scientists who perform statistical research. It was one of the main themes of the 25th annual conference of the Association for Psychological Science that was held in 2013.

It still frequently happens that, a few years after all the initial fuss and attention, the same statistical research proves to be non-replicable and quietly disappears into the wastepaper basket. Anyone relying on statistical research should think twice before drawing any conclusions. As the saying goes, "*It ain't what you don't know that gets you into trouble. It's what you know for sure that just ain't so*".

2 Amazing Coincidences

Defying the Odds

One in a Million

Bizarre things happen. In June 1980 an American woman named Maureen Wilcox bought lottery tickets for both the Massachusetts State Lottery and the Rhode Island Lottery. You can probably imagine her surprise and disbelief when she hit the winning numbers of both lotteries. And yet, her disappointment must have been much greater. Why? The Massachusetts ticket showed the winning numbers of the Rhode Island Lottery and the Rhode Island ticket showed the winning numbers of the Massachusetts Lottery. So instead of hitting the jackpot twice, she won nothing at all.

Life is and always has been full of amazing coincidences. Aristotle already mentions the instance of a farmer who finds a hidden treasure while digging a hole to plant a tree. It has become a classic example in philosophy. Or take for example this modern classic: you're thinking of someone you haven't seen or spoken to for a long time when all of a sudden your smartphone rings. It turns out to be that very same person.

Mathematicians Frederick Mosteller (1916–2006) and Persi Diaconis have collected notebooks and file folders full of striking and unpredictable coincidences, and a few years ago the British statistician David Spiegelhalter started a website under the name *The Cambridge Coincidences Collection*. People can submit all kinds of coincidences they have experienced themselves. In its first year and a half the collection already contained over four thousand coincidences. And these are only the ones people found to be worth telling.

Coincidences fascinate us because they are unpredictable and highly improbable. They take us by surprise and provoke amazement or even disbelief. The most bizarre ones often make headlines.

On February 14, 1986, the *New York Times* released the following news on its opening page: "ODDS-DEFYING JERSEY WOMAN HITS

LOTTERY JACKPOT SECOND TIME. Defying odds in the realm of the preposterous — 1 in 17 trillion — a woman who won $3.9 million in the New Jersey state lottery last October has hit the jackpot again and yesterday laid claim with her fiancé to an additional $1.4 million prize".

On February 26, 2008, English newspapers the *Daily Mail* and the *Sun* published the following story: "Proud Martin and Kim MacKriell never forget their kids' birthday — because all three were born on the SAME date … January 29. Experts calculate the odds of a couple having three children all on the same date are 7.5 in a million!"

And more recently, on February 24, 2016, *ABC News* reported that an American woman, Lindsay Hasz, bit down on a rare purple pearl while eating frutti di mare in her favorite Italian restaurant. According to the president of the Northwest Geological Laboratory, Ted Irwin, the chances of finding a natural, gem-quality pearl like the one Lindsay Hasz found are probably "one in a couple million".

All these coincidences were extremely improbable — "*1 in 17 trillion!*", "*7.5 in a million!*", "*one in a couple million*" — and yet they happened. How can this be explained by chance alone? It certainly looks as if these incredibly improbable coincidences really do defy the odds, as the *New York Times* exclaims.

Winning the Jackpot Twice

Winning the top prize in a lottery, as we saw in the previous chapter, is something many people dream of, but the chances of this dream coming true are extremely slim. Winning the lottery *twice* is so incredibly unlikely that no one even dreams of it. And yet it happened.

In early February 1986, an American woman named Evelyn Adams won the New Jersey state lottery jackpot for the second time in four months. The *New York Times* headlined that the odds were just one in seventeen trillion. That's 1 in 17,000,000,000,000. A number that is simply inconceivable to humans. Did Evelyn really defy the odds?

Let us see what this sensationally small figure actually stands for. On the surface, it appears to be the probability of hitting the jackpot twice after buying two separate tickets for two separate drawings of the New Jersey state lottery. Is this what really happened? The answer is no.

Evelyn had bought multiple lottery tickets every week, and she had been doing so since 1970. There also was a four-month period between the first time she won the lottery and the second time.

To be fair, the lottery officials were quick to revise the statement about Evelyn's chances. Nevertheless, the chances of winning the lottery twice remain extremely slim. To give you an idea, no one had ever hit the jackpot twice in any of the twenty-two state lotteries in the United States before. So how can we explain this exceptionally rare coincidence?

Intuitively, we tend to view surprising coincidences as isolated events. This is very understandable, as coincidences are exceptions to the rule. But this focus blurs our vision of the surrounding landscape. If we want to put these coincidences in their proper perspective, we need to zoom out and look at the whole picture. In other words, we should treat Evelyn as one of the many people who participate in the New Jersey state lottery or a similar lottery.

Before Evelyn Adams won the lottery twice, you had probably never heard of her. She was just one of the many people who bought lottery tickets in the New Jersey state lottery. What if she hadn't been the one who had won twice, but someone named John Smith. Would that have made any difference? Of course it wouldn't have. It would have been just as remarkable.

And why focus on the New Jersey lottery? If it had been the Louisiana state lottery or any of the other twenty-one state lotteries in the United States it would have been equally remarkable. So instead of asking what the chances were that Evelyn Adams would win the top prize twice, we have to ask what the chances are that *somebody somewhere in the United States* will hit the jackpot twice.

Two statisticians from Purdue University in the state of Indiana, Stephen Samuels and George McCabe, calculated this probability. Based on some realistic assumptions, they concluded that for a four-month period the probability of hitting the jackpot twice is 1 in 30 (3.33%) — a far better chance than the one in seventeen trillion chance mentioned in the *New York Times*! For a period of seven years, the same odds even climb to a surprising 50%.

How can we explain this huge increase in probability? The reason is simple. Millions and millions of people in the United States buy lottery tickets, not just one but often several at a time, and often from multiple lotteries. If a large enough number of people participates in lotteries for a long enough period of time, someone is bound to win the lottery twice by chance alone. That someone happened to be Evelyn Adams.

What happened to Evelyn is, in a way, exceptional. Hardly anyone will ever win a lottery twice. But, as the calculations of the Purdue University statisticians show, winning the same lottery twice is not in itself impossible, even though the chances for each individual are close to zero.

Persi Diaconis compared the occurrence of such "one-in-a-million events" to the "blade of grass paradox". Imagine standing in a meadow. There are millions of blades of grass. You reach down to randomly touch one of them. The probability that you end up touching the specific blade you do was extremely small beforehand. One in millions. But the probability that one of the blades would be touched by you was one.

After her exceptional luck in the lottery, Evelyn told the press: "I'm going to quit playing, I'm going to give everyone else a chance". No doubt she was joking, but Lady Luck wasn't amused. She turned the tables on her and, over the next fifteen years, Evelyn Adams wasted her capital to the last penny in the gambling palaces of Atlantic City.

A Big Joint Birthday Party

It is the privilege, or the curse, of twins that they share the same birthday. If you have a brother or sister and you are not twins, the chance of sharing the same birthday is only about 1/365 (0.27%).

For the MacKriell family from the UK January 29, 2008 was a more than special day. Not only was their third child, Ruby, born that day, but also their son Robin and their daughter Rebecca had been born on exactly the same day fourteen and twelve years earlier! Friends insisted on going to the newspaper with this remarkable story. And son Robin already imagined their upcoming birthday party: "like Christmas without a Christmas tree…".

According to statisticians at Cambridge University and at the London School of Economics, the probability of three children in one family being born on the same day is 7.5 in one million. That's an extremely small probability of 0.00075%. So it seems almost impossible that what happened to the MacKriell family was a coincidence. But how did the statisticians come up with such a small figure? It turns out they simply calculated the probability that three children have the same birthday. This is so easy to calculate we can do the calculation ourselves.

Assume that the probability of a child being born is the same for each day of the year and that the births of the three children are independent of each other. Ignore leap days. Now the first child can be born on any day of the year. Since a year has 365 days, the probability that the second child is born on the same day of the year is 1/365. And the same goes for the third child. So the probability that all three have the same birthday is given by the product rule: $1 \times 1/365 \times 1/365 = 1/133{,}225$ which is about 0.00075%. In other words, about 7.5 in one million.

Is this figure really representative of what happened? Once again, the answer is no. To show you why, let's zoom out to see this coincidence in its proper perspective. There are 24 million households in the UK. One million of them, including the MacKriell family, consist of two parents with at least three children. Since the probability of these three children having the same birthday is 7.5 in one million, it is to be expected that in the UK about seven to eight families happen to have their three children born on the same day of the year. In other words, the MacKriell family is most likely not unique at all.

The Law of Truly Large Numbers

Zooming Out

Other highly unlikely coincidences, such as the example of Lindsay Hasz biting on a rare purple pearl while eating frutti di mare in her favorite restaurant, can be explained in the same manner. When you zoom out and look at these coincidences from a broader perspective, they turn out to be pure chance events.

Frederick Mosteller and Persi Diaconis put it this way: "Truly rare events, say events that occur only once in a million, are bound to be plentiful in a population of 250 million people", i.e. in the USA. And zooming out still further: "Going from a year to a lifetime and from the population of the United States to that of the world, we can absolutely be sure that we will see incredibly remarkable events".

The two mathematicians, for understandable reasons, named the principle underlying this the *law of truly large numbers*. In short, the principle states that any miraculous coincidence is likely to happen when a sample or a population is large enough.

In the case of Evelyn Adams, the relevant context was the millions and millions of people in the United States who frequently buy lottery tickets, often several at a time and often from multiple lotteries. In the case of the MacKriell family, it was the one million households in the UK that consist of two parents and at least three children.

Spotting Black Swans

The law can be illuminated with the help of some simple math. Take a one-in-a-million event, say the spotting of a black swan. Imagine there is a lake with a population of one thousand swans. What are your chances of spotting at least one black swan?

The easiest way to calculate this probability is by calculating the complementary probability, the probability that you won't spot a black swan. From this you can easily infer the probability of spotting at least one black swan.

The complementary probability that a swan is not a black swan is $1 - 1/10^6$. According to the product rule, the probability that you do not spot a black swan in a lake of one thousand swans is therefore $(1 - 1/10^6)^{1000} = 0.999$ (99.9%). So your chances of spotting at least one black swan are $1 - 0.999 = 0.001$. That is a small 0.1%.

But what if we extend our observations to a *hundred* lakes, each with a thousand swans? Then the probability of not spotting a black swan shrinks to $(1 - 1/10^6)^{100,000} = 0.905$ (90.5%). And if we observe a *thousand* lakes with a thousand swans each, the probability of not spotting a black swan even drops to $(1 - 1/10^6)^{1,000,000} = 0.368$ (36.8%).

So by amplifying the initial population of one thousand swans first a hundred times and then a thousand times, our chances of spotting a black swan increase from a tiny 0.1% to a significant 9.5% (100% – 90.5%) and a robust 63.2% (100% – 36.8%) respectively.

Clearly, the larger the population of swans, the more difficult it becomes to not spot a black swan. In other words, in a large enough population of swans you are more or less bound to spot a black swan. And this is exactly what the law of truly large numbers tells us: *if a sample or a population is large enough, extremely unlikely coincidences are likely to happen by chance alone*. Such an extremely unlikely coincidence happened in 1954 in a small town called Sylacauga in the state of Alabama.

A Rock from Space

On a sunny afternoon in November 1954, Ann Hodges was taking a nap on the couch when suddenly a piece of black rock crashed down through the ceiling. It bounced off the radio and hit Ann in the thigh. The rock, the size of a grapefruit and weighing about four kilos, gave her a huge bruise. Where did it come from? The surprising answer was: from outer space. Ann Hodges had been hit by a meteorite.

The vast majority of meteorites end up in the ocean or in sparsely populated areas. Astronomer Michael Reynolds emphasized how exceptional this coincidence was: "You have a better chance of getting hit by a tornado and a bolt of lightning and a hurricane all at the same time…". So the odds were incredibly small and yet it happened to Ann.

But if you think of all the centuries that the earth has been exposed to a bombardment of meteors and if you take into account the rapid growth of the world's population in the last two centuries, then you may understand that in accordance with the law of truly large numbers this was almost bound happen somewhere, sometime.

There was no reason why the meteorite would hit Ann. It just happened. Yet the residents of the town who gathered at Ann's house came up with all kinds of scenarios. Some were convinced that a plane had crashed, while others blamed the Soviet Union.

Ann, overwhelmed by the anxious crowd, was taken to hospital, and the head of the local police seized the black rock and handed it over to the Secret Service of the U.S. Air Force. The only one who suspected a meteorite had hit the unfortunate Ann was a local geologist. A few days later, Air Force experts confirmed his opinion. It was indeed a meteorite.

What was to be done with the extraterrestrial rock? Most people thought it should be returned to Ann. And Ann agreed: "I feel like the meteorite is mine. I think God intended it for me. After all, it hit me!" But two years later Ann and her husband decided to donate the meteorite to the Alabama Museum of Natural History at the University of Alabama in Tuscaloosa, where it can still be admired today.

Oracles and Predictive Dreams

Too Amazing To Be A Coincidence

Like Ann Hodges, many people believe that exceptional coincidences have some deeper meaning. They are "too amazing to be a coincidence". Already in ancient times people were fascinated by coincidences that appeared to occur for some reason. In his biography of the Roman general Sertorius, Greek philosopher and historian Plutarch (46–119) mentions that there were people who took pleasure in collecting remarkable coincidences, from reading or hearsay, that appeared to be the result of rational design.

Plutarch himself was skeptical about these claims. He wrote: "since chance continuously and erratically changes her course and time is boundless, one should not be amazed that coincidences often spontaneously occur". The Greek historian explained these unlikely coincidences by the law of truly large numbers, nearly two thousand years before the principle was given its name.

More than any other kind of coincidences, predictions that come true have been and still are considered as too amazing to be pure coincidence. The Stoics believed that the human soul was of divine origin and therefore had privileged access to future destiny. Today, people speak about "extrasensory perception". The ancients spoke of divination and predictive dreams, while parapsychologists now talk about "precognition" and "precognitive dreams".

The basics have remained the same, and so has the critical response. Ancient skeptics like Carneades, Cicero, and Plutarch and modern scientific skeptics all agree: most of these so-called predictions are coincidences that can be accounted for by the law of truly large numbers.

Predictive Dreams about Disasters

Dreams that match future events can be quite impressive and convincing, especially when they deal with some dramatic event. After the sinking of the Titanic in 1910, many premonitory dreams were reported. More recently, people all over the United States claimed they had a precognitive dream in the days leading up to the terrorist attacks of 9/11.

On August 4, 2020, a depot containing a large amount of ammonium nitrate exploded in the port of the Lebanese city of Beirut. The explosion caused 203 deaths and 6,500 injuries and left an estimated 300,000 people homeless.

Imagine that the night before you dreamed of a large explosion and the next day you watch on television the images of collapsed buildings and flames spreading everywhere. It is hard to escape the unsettling feeling that you witnessed the future disaster in your dream.

Many people in the Netherlands had this feeling on May 13, 2000. On that day in the city of Enschede a fireworks explosion destroyed 400 houses, killed 23 people, and injured more than 900. Afterwards, many people reported they had dreamed of an explosion the night before. But as impressive and convincing as such seemingly precognitive dreams may be, they can be accounted for by chance alone.

Let us put this predictive dream in perspective and treat it as one dream among many. As Cicero wrote, people dream a lot. We now know that every night we dream for about two hours and that during this time we have about six dreams. Over a lifetime of 72 years, that adds up to about six years! In total we have around 72 × 365 × 6 dreams. That's more than 150,000 dreams, a truly large number.

Now, how many of these dreams are about an explosion? It's hard to say, but probably not too many — otherwise, there would have been nothing special about dreaming of an explosion the night before the disaster took place. Suppose people dream about an explosion an average of three times in their lives. In other words, only 1 in 50,000 dreams is about an explosion, a very cautious estimate. At first glance this is such a rare event that is hard to believe that having this dream the night before the disaster could be mere coincidence. Yet the law of truly large numbers tells a different story.

Throughout the Netherlands, with an adult population of about 15 million people at the time of the explosion, a total of about 90 (15 × 6) million dreams were dreamed every night. Based on the assumption that 1 in every 50,000 dreams is about an explosion, we may expect that the number of people who had a dream about an explosion the very night before the disaster was equal to 90 million divided by 50,000, i.e. 1,800 people.

Say 3 out of 4 people immediately forgot their dream, then 450 people would still be left who had a "predictive dream" the night before and presumably about four of them even lived in the city of Enschede itself, given the size of its population. So, as impressive as these dreams may be, they are not precognitions of future events, but merely blind matches that occur by chance alone.

Ancient Oracles

Another ancient manifestation of foreseeing future events, one that still appeals to our imagination, was the oracle. On the slope of Mount Parnassus lay the Greek city of Delphi. According to the Greeks, it was the center of the earth. People from all over the world visited the city. From peasants to kings, everyone came to consult the Pythia.

The Pythia was the high priestess of the Temple of Apollo and the most important oracle of Greek Antiquity. She was consulted about all kinds of future events, from the end of a love affair to the receiving of an inheritance. The most famous was the consultation by Croesus, the king of Lydia.

According to the Greek historian Herodotus, the king asked the Pythia if he should continue his campaign against Persia. The oracle replied that if he attacked the Persians, he would destroy a great empire. Convinced that the answer was favorable, Croesus invaded Persia. Unfortunately, he was defeated by the Persians. The great empire that was destroyed was his.

Vagueness and enigmatic ambiguity are typical of the predictions of ancient oracles. Whatever happens, the oracle cannot be mistaken. It is your interpretation that is right or wrong. Quite different were the straightforward predictions of the most famous oracle of recent times: a cephalopod named Paul.

Paul the Octopus. A Modern Oracle

The superstar of the 2010 soccer World Cup in South Africa was not a team or player, but a cephalopod who spent his days in a tank at the Oberhausen Sea Life Centre in Germany. The animal became a worldwide celebrity because of his extraordinary ability to pick the winning team before a match was played.

Paul the Octopus, as he was called in the newspapers and on social media, predicted the winning team of all seven matches played by the German national team, including the lost semifinal against Spain. Moreover, he correctly predicted Spain as the winner of the World Cup final against the Netherlands.

How could a cephalopod predict the outcomes of soccer matches? Prior to each match two transparent, closed boxes marked with the flag of one of the two opposing teams were placed in front of Paul. Each box contained a mussel, as Paul was a real mussel lover. The winner of the match would be the team whose flag was on the box that Paul opened first. He correctly predicted the winner eight matches in a row.

The octopus received worldwide media attention after he correctly picked the winner in a series of four matches. Even serious newspapers like the *Washington Post* and the French newspaper *Le Monde* began to report on the psychic powers of the eight-armed oracle. The day before the final took place over two hundred journalists were present at the Oberhausen Sea Life Centre, including several television crews from different countries.

Soccer fans all over the world were captivated by the octopus's predictions. Paul was exceedingly popular, particularly in Germany. That all changed when the octopus correctly predicted Germany's defeat in the semifinals against Spain. Paul received death threats and many German fans expressed the wish that he would end up on a dinner menu.

In Spain, on the other hand, people were extremely enthusiastic. In response to the many threats, Prime Minister José Zapatero offered Paul state protection. Two cabinet ministers called for the octopus to be granted asylum.

All the commotion presumably went completely unnoticed by the animal itself. At the height of his fame, after predicting the winner of the 2010 World Cup final, Paul the Octopus went into retirement.

Octopuses are among the most intelligent invertebrates on earth. They can unscrew bottles and are even capable of facial recognition. Paul used to look back at visitors when they came near his tank. According to the director of the Sea Life Centre, Paul showed signs of a special intelligence early on in his life. Could Paul the Octopus, unlike his peers, have been a soccer fan with an astonishing understanding of the game?

It would be highly unlikely for an underwater invertebrate, but, on the other hand, his remarkable performance seems too amazing to be mere coincidence. If we assume that he had an even chance of correctly predicting the winner of each match, then the probability that he would get it right eight times in a row equals $(1/2)^8 = 1/256$. Less than 0.4%. It appears as if the octopus really had the power of prophecy.

Mani the Parakeet and Other Oracle Animals

Paul the Octopus' predictions become a lot less spectacular if we put them into perspective. For all the media commotion about the psychic octopus, he was by no means the only animal on earth that was consulted about the World Cup matches. Everywhere in the world animals were and are regularly consulted, not only about sports results, but also about stock market prices and other phenomena that are difficult to predict.

During the World Cup, Paul competed with animals such as Harry the Crocodile in Australia, Pino the Chimpanzee in Estonia, and Pauline the Octopus in the Netherlands. His biggest rival in the media, especially in Asia, was the parakeet Mani in Singapore. This psychic bird had previously assisted his master Muniyappan, an 80-year-old Indian fortuneteller, in choosing lottery numbers and in picking the best date for couples to get married.

However, Mani quickly vanished into public oblivion when he was less lucky and predicted that the Netherlands would win the finals. Little to nothing is heard about these and many other animals that were less successful than Paul. It's like watching a baseball game on television and all you get to see are the home runs. What if not Paul the Octopus but the parakeet Mani had correctly predicted the winner of the World Cup final and had made headlines everywhere? We'd be just as amazed, but now about a clairvoyant parakeet.

Let us put the performance of the cephalopod into perspective and treat Paul the Octopus as just one of these many "psychic" animals. Suppose each of the 32 countries in the 2010 World Cup play-offs had its own national animal that predicted the winning team of a match, and let's calculate the probability that at least one of them would correctly predict the winner of eight matches in a row. This is done in exactly the same way as in the case of spotting a black swan.

Suppose that for each match the animals have an even chance of predicting the winner, like flipping a coin. For each animal, the probability of correctly predicting the winner of eight matches in a row is $(1/2)^8 = 1/256$. So the complementary probability that none of these 32 animals will succeed in predicting a winner in eight consecutive matches is $(1 - 1/256)^{32} = 0.882$ (88.2%). This means that the probability that at least one of them will correctly predict eight winners in a row is about 12% — a modest but substantial chance, far from being "too amazing to be a coincidence".

Octopuses usually don't live long, even if they don't end up on the menu. A few months after the World Cup, in October 2010, Paul died peacefully in his tank at the Sea Life Centre in Oberhausen. His ashes are kept in a memorial in the form of a massive octopus spreading his tentacles around an over-sized soccer ball. It is still on view for the public.

A Modern Miracle

Miracles

"Miracle thinking" is wired into our brains. The spotlight of our intuition has a limited reach; it focuses on everything that is extraordinary if viewed from up close, but may not be so exceptional from a panoramic point of view. And even if our intuition zooms out to a wider perspective, it doesn't take into account all cases where something might happen, but actually doesn't. We would not know where to look, and if we did, we'd just freeze!

So we have no clue how small the probability is that a coincidence does not happen at least once, somewhere sometime. And as a result, we are amazed when a seemingly exceptional coincidence occurs. It may even feel like a miraculous event. Yet most of these "miracles" can be explained by the law of truly large numbers. As Cicero was wise enough to note about two thousand years ago, chance is responsible for all kinds of miracles throughout the ages.

You may be wondering whether there are any coincidences left that are so unlikely they will never occur, at least not by chance alone. The answer is: yes, there are. How about a million people tossing a coin and all the coins landing heads up? Or what about the famous monkey who randomly hits a keyboard and types out the first line of Shakespeare's Hamlet? If any of these things ever happened, it would really be a miracle. Like the following true story, for instance.

The Miracle of Kineton

In the last days of November 2011, something exceptional occurred in the village hall of the English town of Kineton in southeast Warwickshire. Something that certainly seemed like a miracle, at least in the perception of the four retirees preparing for a game of whist, an English card game and forerunner of the popular card game bridge.

The four friends had come to the village hall on their regular evening and were preparing for the first game. The deck was shuffled and cut, and the cards were dealt as usual, with each player being dealt a hand of thirteen cards. Wenda Douthwaite, the 77-year-old former owner of a local tearoom, was the first to pick up her hand of cards.

One can hardly imagine her astonishment and disbelief. She held a complete set of spades in her hand! But that was not all. "We compared cards and were totally shocked. I was shaking when we lay the cards down on the table", Wenda said. The others were dealt complete suits as well. "We've never seen anything like it before".

This once-in-a-lifetime event made the newspapers all over the world. On November 24, 2011, the British newspaper the *Sun* headlined: "Card players dealt one in two thousand quadrillion hand". And the *Daily Mirror* reported: "Four pensioners playing a game of whist couldn't believe their eyes when they were each dealt a complete suit of cards — an almost impossible feat".

A player's hand with all thirteen cards of the same suit — hearts, diamonds, clubs, or spades — is known as the *perfect hand*. Occasionally, newspapers report that a bridge player has been dealt a perfect hand, but to date, no official tournament or championship anywhere in the world has ever witnessed anyone getting a perfect hand.

So what about this "almost impossible feat" of all *four* players getting a perfect hand? Could this really ever happen, as allegedly was the case in the small English village of Kineton? According to the newspapers, the probability of it happening is one in two thousand quadrillion. A quadrillion is a one with 24 zeros, an astronomical figure. To put this outrageous coincidence into perspective, we will leave Earth and travel to a small but extraordinary planet far away that is inhabited by living creatures obsessed with playing bridge.

Planet Bridge

Far from Earth, a small planet inhabited by humanoid beings revolves around a different sun. The days on the planet are shorter, but just like on

Earth, a year has 365 days. The planet, with a pleasant subtropical climate, consists mainly of one long coastline with wide sandy beaches and waving palm trees. Food is plentiful and all inhabitants form happy couples for life.

As it happens, very little happens on this planet. Except playing bridge. The extraterrestrials play bridge every day without getting bored. It's pretty much all they know. What's more, the number of different possible deals is larger than the number of stars in the universe. So there's little chance that these amazing bridge fanatics will ever get bored by playing the same game twice.

The small planet happens to have exactly as many adult inhabitants as people on earth who play bridge, about 50 million. Since every couple on the planet forms a bridge pair, there are 25 million pairs of players. Each couple has the habit of playing ten games every morning. Then they eat on the beach and enjoy a long siesta in the shade of a palm tree. At the end of the afternoon, after the heat has subsided, the couples play another ten games of bridge, eat a small dinner, and go to sleep. That's pretty much how they spend their time, day in and day out, year after year.

The Last Deal

The extraterrestrials are very relaxed and almost fearless. The one thing they fear is a combination of cards they call the *Last Deal.* It is so special and so rare that they believe the world will end when it is dealt. And that is when all four players get a perfect hand. Do these distant bridge fanatics really have to worry that the end of time is near?

If we want to know how real this threat of doomsday is, we first need to figure out the probability of all four players getting a perfect hand. That is, the number of ways all four players can get a hand of one suit divided by the total number of ways the cards can be dealt.

The number of different ways in which the four suits can be dealt to the four players is $4 \times 3 \times 2 \times 1 = 24$. (In mathematics this number is called the *factorial* of 4 and notated as 4!) Player one can get any suit out of the 4 possible suits, player two any suit out of the 3 remaining suits, player three any suit out of the 2 remaining suits and player four the 1 suit that is left. So the four suits can be arranged among four hands in 24 different ways.

Furthermore, since a suit consists of 13 cards, each suit can itself be dealt in $13 \times 12 \times 11 \times \ldots \times 3 \times 2 \times 1 = 13!$ different ways. So the total number of ways in which four perfect hands can be dealt is $24 \times (13!)^4$.

This is quite an impressive number, but it is nothing compared to the total number of possible deals. This number is equal to $52 \times 51 \times 50 \times \ldots \times 3 \times 2 \times 1 = 52!$ For the first card that is dealt, there are 52 possibilities, for the second card 51 and so on.

The probability of being dealt the Last Deal is the number of favorable outcomes divided by the total number of possible outcomes, that is $24 \times (13!)^4/52!$ This is an incredibly small number, $4.47/10^{28}$. It is the "one in two thousand quadrillion" chance that the *Sun* mentioned.

What about the probability that the planet will perish in the short term? In other words, what is the probability that the Last Deal will be dealt in, say, one year? We know there are 25 million bridge pairs who play bridge, with each playing 10 games every morning and another 10 games at the end of the afternoon. The number of games that are played every day therefore equals $(25{,}000{,}000 \times 20)/2 = 250{,}000{,}000$. We have to divide by two, otherwise each game would count double. Since a year has 365 days, the total number of games played each year is $250{,}000{,}000 \times 365 = 91{,}250{,}000{,}000$.

The probability that the Last Deal will be dealt in a year can be determined by first calculating the complementary probability that no Last Deal will be dealt. For one game this probability is $1 - 4.47/10^{28}$. In other words, the probability that this won't happen in any of the more than 91 billion games played in a year is equal to $(1 - 4.47/10^{28})^{91{,}250{,}000{,}000}$.

If you subtract this number from 1, you'll get the probability that the Last Deal *will be dealt* in one year. This probability turns out to be about $4{,}1/10^{17}$, which is 0.0000000000000041%. In other words, practically zero.

What if we take a hundred years instead of one year? It's easy to mathematically prove that the probability increases in proportion to the number of games as long as the probability remains very small. So the probability will be about a hundred times greater in a hundred years' time. In other words, still incredibly small.

In short, even these extraterrestrial bridge fanatics will never experience four perfect hands in their lifetimes. Only in a time span of 17,000,000,000,000,000 years will they have a 50% chance of being dealt the Last Deal, which is far beyond the expiration date of the known universe!

If it is virtually certain that this alien civilization of 50 million bridge fanatics will never witness four perfect hands in the same deal, it is even more certain that their 50 million earthly counterparts never will.

Kineton Revisited

What about the miracle that happened in Kineton? If it wasn't a practical joke, the most likely explanation is that the cards weren't shuffled properly. The mathematician Persi Diaconis, who began his career as a magician, has proven that a deck of 52 cards must be thoroughly shuffled at least seven times to ensure that the cards in the deck are in random order. If not, as often happens, the probability of getting four perfect hands greatly increases. In that case the miracle of Kineton can simply be accounted for by the law of truly large numbers.

The Birthday Paradox

A Mathematician's Restaurant

Imagine that in the center of the city you live in a modest but remarkable restaurant has recently opened its doors. Remarkable, because the owner is a mathematician. As a mathematician, he had reached the peak of his powers. Thinking about how he wanted to spend the rest of his life, he decided to follow the example of his favorite opera composer, Gioachino Rossini. He quit his profession to devote himself to his second greatest passion, gastronomy.

Thanks to some peculiarities he inherited from his mathematical past, his restaurant soon became a trendy nightlife venue. The dishes were prepared with mathematical precision and elegance. They were given names like *Archimedean Spiral*, *Pascal's Triangle*, and similar names. And from time to time he presented his guests with a mathematical riddle. Whoever solved it first won a free dinner.

One day you decide to visit this intriguing restaurant. It happens to be the owner's birthday. The small restaurant is completely booked with 23 guests, including yourself. Because today is his birthday, the host proposes the following bet to his guests. If two or more guests have the same birthday (not necessarily the host's birthday), the lucky ones get a free dinner on the evening of their birthday. However, if none of the guests happen to share the same birthday, they will have to leave a special birthday tip.

Some enthusiastic guests immediately start exchanging the dates of their birthdays, while others are not so enthusiastic and start protesting. The bet isn't fair. There are 365 possible birthdays and only 23 guests! You are in doubt. The protesters seem to have a point. On the other hand, why would the host, on his own birthday, propose an unfair bet? The only way to find out is to calculate the odds. But how do you tackle this problem? Using a famous example, we'll first show you how not to do so.

And now… here's Johnny!

We leave the restaurant for a moment, and travel back in time. It is the evening of February 6, 1980 and the *Tonight Show* is on American television, hosted by the unbeatable King of Late Night television, Johnny Carson (1925–2005).

Johnny Carson, for those unfamiliar with him, is one of the great icons of American television history. From 1962 to 1992 he hosted his talk show five nights a week. He welcomed numerous celebrities and introduced the world to newcomers such as Steve Martin, Jim Carrey, Madonna, and Ellen DeGeneres. Everyone wanted to appear on *The Tonight Show Starring Johnny Carson*.

On tonight's show, Johnny delivers one of his hilarious monologues in which he addresses the "birthday problem" as it is called in mathematics. Outside mathematics, it is commonly known as the "birthday paradox". After some bantering about the upcoming birthday of his regular sidekick, Ed McMahon, Johnny addresses the audience.

How many people in the audience have a birthday in February, just like Ed? Some of the audience clap their hands. Then Carson tries something he remembers from a long time ago. After some difficulty finding the right words, Johnny comes up with his own memorable formulation of the birthday paradox.

Johnny: *"How many people would you think would have to be in a room that your odds would be almost sure that they would have exactly the same birthday on the same day?"*
(Johnny means: *how many people have to be in a room to be almost 100% certain that two of them have their birthdays on the same day?*)
Ed guesses: *"One thousand"*. It's obvious Ed doesn't quite understand the question.
Johnny Carson shakes his head from no and says: *"Something like 35 or 40."*
Ed interrupts, surprised: *"That's all?"*
Johnny continues: *"The odds are pretty good, if I remember."*
Ed to Johnny: *"Well, pick a day and see if we have it. Just pick a date."*
Johnny agrees: *"Well, let's just check this audience here."*

Johnny Carson turns to the audience: *"Lady in the front row, what's your birthday?"*
The woman answers: *"August 9th!"*
Johnny addresses the audience: *"Anybody else here have a birthday on August 9th?"*
Johnny looks astonished, not knowing what to say: *"Nobody? ... We have 500 people there, nobody has August the 9th? ... Ok, we missed there. All right, somebody else?"*
Someone in the audience shouts out loud: *"April 9th!"*
Ed: *"Anybody else here have a birthday on April the 9th?"*
No one answers. The audience, finding things are getting hilarious, starts laughing. Ed smiles triumphantly, while Johnny reacts angrily to hide his disappointment.
Johnny: *"Look, I may have had my figures a little bit wrong."*
Ed, who is feeling more and more confident, starts a discussion. Of course it is impossible, he argues, that among 35 people two have their birthday on the same day, because there are 365 days in a year!
But Johnny answers: *"No, I know it doesn't sound possible ... but the law of probabilities still ..."*
And then to the audience: *"All right, now we're going try something — forget you two dummies who obviously are making up phony birthdays. All right, let's try mine. October 23rd."*
Ed: *"Anybody?"*
This time Johnny gets the response he's looking for. Two people in the audience raise their hands. Johnny seems to be saved in the end.

Carson's Mistakes

Did Johnny Carson really prove that in a room with 35 or 40 people it is almost certain that two of them have the same birthday? Of course not! First of all, the numbers he mentions don't add up. For 35 to 40 people there is a pretty big chance, but it is not at all certain that two of them share the same birthday. Johnny's memory is probably failing him, as he begins to suspect during the show.

Secondly, there are no fewer than five hundred people in the audience — not exactly representative of a group of only 35 or 40 people. Of course, in an audience of five hundred it is 100% certain that at least two people will share the same birthday. After all, the number of people in the audience exceeds the number of days in a year by 135.

However, these mistakes are less important than the crucial mistake made by Johnny, a mistake that many people make when they first hear about the birthday paradox. Johnny was looking for someone with the same birthday as a *specific* other person, in this case the woman in the front row. Or, later, Johnny himself. But the birthday paradox is about *any two people* in the room having the same birthday.

These are two totally different cases, each with a totally different probability. Let's call two people who share the same birthday *a birthday couple*. Johnny confuses the number of *all* possible birthday couples with the number of possible birthday couples that can be made with one *specific* person.

One way to get out of this confusion is to follow the same procedure we applied previously. If two people in a group have the same birthday, the probability of this birthday couple is only 1/365, which is 0.27% — at first glance, an exceptional coincidence. However, we should not focus on these two people, but put the birthday couple in its proper perspective, that is, we should zoom out to all possible birthday couples in the group. In other words, we should be looking for the probability that *any two members of a group* share the same birthday.

Before we leave Johnny Carson and return to the present, we have to admit that Johnny also had some bad luck. The probability that in an audience of five hundred people someone shares his or her birthday with a *selected* person is about 75% (as can easily be calculated by first calculating the complementary probability that no one in the audience shares this birthday). The probability of missing the mark twice in a row, as Johnny Carson did, is only about 6.4%.

Amazing Odds

Imagine you're back in the restaurant, together with the other 22 guests. Intuitively, you choose the side of Ed and some of the guests. Since there are 365 different possible birthdays, it doesn't seem very likely that two of the 23 guests will have the same birthday. The only way to know for sure is to calculate this probability. But how? The number of people and the number of possible different birthdays make your head spin. Fortunately, there is a shortcut: first calculate the complementary probability.

The opposite of at least two people having the same birthday is nobody having the same birthday. So the complementary probability is the probability that none of the 23 people shares the same birthday. Let's calculate this complementary probability.

For simplicity's sake, we won't take leap years into account. We will also assume that the guests' birthdays are unrelated, so there are no twins among the guests. And we will not take into account the seasonal and weekly fluctuations of the birthrate during the year. By the way, if we were to, the probability slightly increases, but that increase is almost negligible.

Let's start with a simple example. Suppose there are only three guests in the restaurant. The total number of possible combinations of three birthdays would then be $365 \times 365 \times 365$ (in mathematics we notate this number as 365^3 and call it the third power of 365).

If none of these guests have the same birthday, the number of possible combinations is as follows. Just choose one of the three guests. He or she can have a birthday on any of the 365 days. For guest number two there are 364 possible days left and for guest number three 363 possible days. In other words, there are $365 \times 364 \times 363$ possible combinations of three birthdays where no one has the same birthday.

Now we only have to divide this number by the total number of possible combinations to get the probability that nobody shares the same birthday: $(365 \times 364 \times 363)/365^3$ (99.2%). So the probability that at least two of the three guests share the same birthday is $1 - 0.992 = 0.008$ small 0.8%.

How about the 23 guests in the restaurant? The procedure is exactly the same. Each guest can have his or her birthday on any one of the 365 days of the year. So the total number of possible combinations of 23 birthdays is 365^{23}.

The number of possible combinations if none of the guests has the same birthday can be determined as follows. Just choose one of the 23 guests. He or she can have a birthday on any one of 365 days. For guest number two 364 possible birthdays remain, for guest number three 363 possible birthdays, and so on until there are 343 possible birthdays for guest twenty-three. So if nobody has the same birthday, there are 365 × 364 × 363 × … × 343 possible combinations of 23 birthdays.

The complementary probability is this number divided by the total number of possible combinations of 23 birthdays: $(365 \times 364 \times 363 \times \ldots \times 343)/365^{23} = 0.493$. If you subtract this complementary probability from 1 you will get the probability we are looking for: $1 - 0.493 = 0.507$ (50.7%). So there is a better than even chance that two or more guests will share the same birthday!

In short, the bet proposed by the mathematical *chef de cuisine* turns out to be a fair bet, especially if we assume that the cost of two free dinners outweighs the total amount of birthday tips.

A Paradox? Not Really

It's surprising that it only takes 23 people to have a better than even chance. If we increase the number of guests, this probability even increases remarkably fast to 1. If you happen to have 57 or more friends on Facebook, the chance that at least two of them share the same birthday is already greater than 99%.

Many people call these counterintuitive outcomes paradoxical, hence the birthday problem is commonly known as the *birthday paradox*. However, there is nothing paradoxical about the birthday problem if you put things in their proper perspective.

Intuitively, we compare the number of 23 individuals with the number of 365 possible birthdays. But the problem is not about individuals, but about pairs of individuals. Birthday couples, as we called them. When we zoom out, the bigger picture will not be a group of individuals but the total of all possible birthday couples. So the actual question is: how many possible birthday couples does it take to have an even chance that at least one of them is a match?

Let's see how many birthday couples — for instance Paul and Irene, Ed and Johnny, Johnny and Irene, and so on — can be made up of 23 individuals. This is easy to calculate. Each of the 23 guests can form a couple with 22 other guests. So the number of possible birthday couples is 23 × 22 divided by 2. You have to divide by two because otherwise you would include all repeat pairs, like for example Paul and Irene but also Irene and Paul. The answer is: 253. So there are 253 possible birthday couples that can result in a match, each with a probability of 1/365 (0.27%).

In short, the birthday "paradox" states that there is a better than even chance that at least one of the 253 possible birthday couples will be a match. This does not sound very paradoxical anymore. Instead, it agrees with what we would expect on the basis of the law of truly large numbers.

We are misled because we tend to greatly underestimate the number of combinations that are possible, even among small numbers of objects or persons. Take the menu of our remarkable restaurant. The menu options are modest. You can choose from four starters, six main courses, and three desserts. All dishes combine well with each other. Say, every week you select a different set of menu items. How many three-course dinners can you put together before you start repeating? The number of different menus is no less than $4 \times 6 \times 3 = 72$. So you can enjoy a different menu every week for more than a year!

The iPod Shuffle

The birthday paradox is not only a popular brainteaser, it is a classic in probability theory. It appears regularly in many different forms, as the following true story illustrates.

On January 11, 2005, a special model of the iPod appeared on the market: the iPod Shuffle. The model had a shuffle function that randomly selected songs from your collection of mp3s and generated a shuffled playlist. However, this wasn't the way it appeared to many users.

Apple received a steady stream of complaints. When people were listening, their iPods seemed to have the tendency to play songs from the same artist or same album or even repeat the very same song in short succession. And that while there were more than a thousand different mp3s on someone's iPod. It's almost as if the shuffle function was picking its own favorite songs!

In reality, the shuffle function was working correctly. Coincidental clusters of tracks from the same artist or the same album gave listeners the impression that the device was not randomly selecting the songs. This impression was even greater when the same song was played a second time not long after the first time. Steve Jobs announced that the shuffle function would be adjusted to avoid these random clusters. "We're making it less random to make it feel more random!"

In fact, the iPod's problem is the birthday problem in disguise. The number of songs in your mp3 collection corresponds to the number of days in a year. The number of tracks you listen to corresponds to the number of people in a room. The probability that two of these tracks are the same song corresponds to the probability that two of the people in the room have the same birthday.

Suppose you have a thousand mp3s on your iPod. In that case the iPod Shuffle problem is equal to the birthday problem on a planet where a year isn't 365 days, but a thousand. How many tracks need to be played to have about a 50% chance of hearing the same song twice?

You'll have to use a pretty good calculator to be able to handle the large numbers. Fortunately, there is a simple approximation formula. That is, a relatively easy formula that gives a good approximation of the correct answer in case it is too difficult or too laborious to calculate the exact answer. The surprising answer is that you only have to listen to 37 tracks to have a 50% chance of hearing the same song twice!

So it's not amazing that many people had the impression that their iPod's shuffle function wasn't working properly. If we include hearing two songs from the same album or by the same artist(s) shortly after each another, this number of tracks will even be a lot smaller.

At first glance, with a thousand different songs, this all seems pretty unlikely. But if you take into account the number of possible combinations of two tracks out of 37, things may appear different to you. Each of the 37 tracks can be combined with 36 other tracks, and if we divide this by two, to exclude repeat pairs, there are as many as 666 possible matches, each with a probability of 1/1000.

In short, poor intuition regarding the erratic behavior of chance, not a software problem, puts us on the wrong track, so to speak. Chance events tend to cluster. Just as long runs of the same outcome are characteristic for the capricious behavior of chance, so are clusters too.

3 Test, Test, Test

Lifting the Fog

A Positive Test for a New Disease

Recently, a previously unknown disease has spread all around the globe. In your country alone, tens of thousands of people — medical experts estimate about 1% of the total population — are already infected. Symptoms don't show until fifteen days after infection. The course of the disease can be very serious. No cure has yet been found.

Fortunately, scientists have succeeded in developing a test that detects 95% of the people carrying the disease, even when symptoms are not yet visible. If someone is not infected, the test will also show this 95% of the time. The authorities have announced that once the test is available on a wider scale, the entire population will be screened for the disease. Those who are infected will be identified and isolated, and in this way the spread of the new disease will be stopped at an early stage.

You feel relieved. There may be no treatment, but at least the authorities will make sure that the disease is brought under control quickly and successfully.

A week later, everyone in the country has been screened, including yourself. You receive a phone call from the municipal hospital. You have tested positive. In other words, you carry the disease! You can hardly believe it. How can this be? You're not the only one who's shocked. When the doctors examine the results of the screening, they are astounded as well. What's going on?

Deep inside you there's a glimmer of hope. Maybe the test is wrong. But that hope evaporates quickly when you recall that 95% of people who aren't infected test negative. That would mean that you have a 5% chance of testing positive and not being infected. It is a depressing thought ... if it were true.

Confusion of the Inverse

The above probabilities are of a different order than the probabilities we have been dealing with so far. *If* you are not infected, *then* the probability of testing positive is 5%. Probabilities that can be formulated in this way are called *conditional probabilities*. A simple example of a conditional probability is the probability that a girl is left-handed: *If* a child is a girl, *then* the probability that she is left-handed is 10%.

Conditional probabilities are notorious, even among mathematicians. A conditional probability and its *inverse* can look very much the same. The probability of testing positive *if you are not infected* closely resembles the inverse probability of not being infected *if you test positive*. The former we know; it's the latter we want to know. But the longer you stare at the two, the harder it is to tell them apart.

Intuitively, people are often inclined to treat a conditional probability and its inverse as the same, or almost the same. But are they really that close to each other? The probability that *a girl is left-handed* is not the same as the probability that *a left-handed child is a girl*. The probability of the former is about 10%, the inverse probability is almost 50% (slightly more boys than girls are left-handed). So both probabilities don't have to be the same at all. The confusion of a conditional probability with its inverse is called the *fallacy of the transposed conditional*.

So what about the probability that the positive outcome of your test is wrong? The probability of testing positive *if you are not infected* is 5%. But that's not what you want to know. What you do want to know is the inverse, the probability of not being infected *if you test positive*. Can doctors tell these two conditional probabilities apart? And if they can, how do they calculate the unknown inverse probability, the probability of not being infected although you tested positive?

Bayes' Rule

Thomas Bayes (1702–61) was an English mathematician and Presbyterian minister who during his lifetime was principally known as a defender of Newton's infinitesimal calculus. It wasn't until later in his life that he began to take a deep interest in probability but he never published anything on the subject himself. Only after his death his notes on probability were edited and published under the title *Essay Towards Solving a Problem in the Doctrine of Chances* (1763).

In the essay he discusses a specific case of the theorem that would make him famous: *Bayes' theorem*, also known as *Bayes' rule*. The theorem states exactly how conditional probabilities and their inverse relate to one another. Would doctors know how to apply Bayes' rule?

A Math Test for Doctors

In 1982 the American physician and mathematician David Eddy (1941–) asked exactly the same question. To find out, he took mammography statistics and put them to the test.

Mammography is still widely used to detect early signs of breast cancer by taking an X-ray picture of the breast, called a mammogram. If a woman has breast cancer, the probability that the mammogram is positive is 79.2%. If she does not have breast cancer, the probability that the mammogram is negative is 90.4%. Since about 1% of all women aged 40 and older suffer from breast cancer, women in that age group are advised to have regular mammography screenings.

Eddy presented these numbers to a hundred doctors and asked each of them what the probability was that a woman, *if* she tested positive, actually had breast cancer. Ninety-five of the hundred doctors estimated the probability at about 75%. They simply assumed that the probability was approximately the same as the 79.2% probability of a positive mammogram *if* the patient had breast cancer. With a few exceptions, they all fell victim to the fallacy of the transposed conditional. Almost none of the doctors applied Bayes' rule.

If the doctors had applied Bayes' rule, they would have seen that the two probabilities are nowhere near each other. The probability of having breast cancer if you test positive is far less than 75%. Yet most doctors didn't know how to deal with conditional probabilities, and as a result they had no clear idea how reliable mammography actually was.

Eddy concluded that in the medical world the application of Bayes' rule was "sporadic and had not yet filtered down to affect the thinking of most practitioners". Eddy hoped that a new generation of doctors, trained in the use of Bayes' rule, would become more aware of the mathematical intricacies of medical testing.

"I'm not a math person"

Eddy's hopes proved to be false. In reality, Bayes' rule is simply too mathematical to be part of a doctor's standard toolkit. To understand Bayes' rule, you must be familiar with some abstract probability theory and its symbolism. But many doctors feel uncomfortable when confronted with the formulas and symbols of mathematics. When faced with a mathematical problem, they try to avoid it rather than tackle it. "I'm just not a math person" is the standard excuse. Why is this "I'm bad at math" such a widespread mantra? The main reason is math anxiety.

People often experience stress when trying to solve a mathematical problem. Even the mere thought of mathematics can be enough to provoke negative feelings. Imagine taking a math class. You have a math problem to solve, but the more you puzzle, the more puzzled you become. You start to feel restless. Your hands get clammy, your stomach gets tighter and you can hear your heart beating. Your mind becomes clouded and you can't concentrate any longer. You just can't find a way out!

This is a basic anxiety. Feeling trapped and not being able to find a way out. The ancient Greeks called it *aporia*. It literally means "no way out". Even mathematicians sometimes experience it. As a result, many people have negative associations with doing math and try to avoid it. Doctors are no exception.

Can we fight these negative associations? Yes, we can. We can do so by showing that solving math problems doesn't have to be traumatizing. Actually, doctors don't need Bayes' rule at all to deal with the conditional probabilities of medical tests. There is a way out that doesn't require mathematical training in probability theory — a method that shows that doing some simple math can already be enlightening and even fun.

Lifting the Fog

In the mid-1990s the German cognitive psychologist Gerd Gigerenzer presented 48 experienced doctors in Munich with four problems about medical tests. The first problem was the same one that Eddy had used, the mammography problem.

About 1% of all German women over 40 suffer from breast cancer. If a woman has breast cancer, the mammogram will be positive with a probability of 80%. If she does not have breast cancer, the mammogram will be negative with a probability of 90%.

Each of the 48 doctors was asked how likely it is that a woman who tests positive actually has breast cancer. One of the doctors — let's call him Dr. A — was an experienced dermatologist and the director of a university clinic.

Although Dr. A was given a blank sheet on which to do his calculations, he used his own piece of paper instead. For ten minutes, he nervously puzzled with the percentages. In the end, by adding the 80% true positives to the 10% false positives, he estimated the probability at 90%. But then he exclaimed in despair, "Oh, what nonsense! I can't do it. You should test my daughter, she is studying medicine".

Clearly, the doctor was in a state of mathematical aporia. He nervously searched for a solution but couldn't find a way out. Gigerenzer explained the same problem to him again, but this time the psychologist didn't use percentages but numbers.

On average 100 out of 10,000 women (1%) suffer from breast cancer. Of the 100 women with breast cancer an average of 80 women (80%) will test positive. Of the 10,000 – 100 = 9900 women without breast cancer an average of 90% will test negative and so an average of 10% will test positive. So there will be about 990 false positives. Again, how likely is it that a woman who tests positive actually has breast cancer?

Once the problem was presented in this way the deep frown on the doctor's forehead disappeared like snow in the sun. "That's so easy", he said, and he solved the problem in no time. All he had to do was to divide the number of true positives (80) by the total number of positive results, i.e. true and false positives (80 + 990 = 1070). Once the problem was restated in numbers instead of percentages, all the mathematical fog lifted!

Using Natural Frequencies

In most traditional medical and statistical textbooks conditional probabilities are presented as a value between 0 and 1 or as a percentage between 0% and 100%. Gigerenzer proposed to express conditional probabilities in a more natural and less technical way. All medical tests should be reformulated in terms of numbers.

For example, instead of saying that 87.5% of all the people who work in the tobacco factory still smoke, you could also say that 175 out of 200 still smoke. Gigerenzer called this the *frequency format*. Because, unlike the percentages and probability values in textbooks, numbers are a natural way of expressing the prevalence of an event, he called these number frequencies *natural frequencies*.

When the mammography problem was restated in natural frequencies, most of the forty-eight doctors were able to solve the problem without any difficulty. One of them even exclaimed: "A first grader could do that. Wow, if someone couldn't solve this ...!" Gigerenzer himself even compared switching to natural frequencies to replacing Roman numerals with modern decimal notation. Both changes hugely simplified calculations.

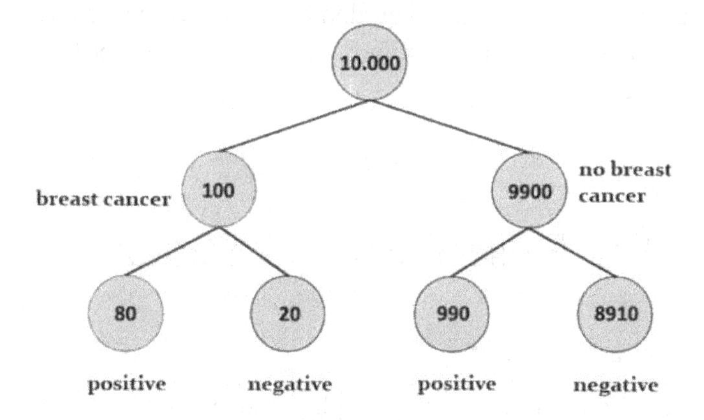

Presentation is everything. Once the problem is reformulated in the frequency format the fog lifts and the problem becomes transparent. If you put the numbers in a tree diagram you can almost "see" the correct solution! In short, it's the natural-frequency approach rather than Bayes' rule that should be in every doctor's toolkit.

What's more, because the frequency approach tackles the problem at its base, that is, the base rate of the disease, you will also no longer fall victim to the fallacy of the transposed conditional.

The fallacy of the transposed conditional ignores the base rate. Why is that? It is because our intuition uses a shortcut. Take the following example. *If your pet is a cat, it probably has four legs*. No one will confuse this conditional probability with its inverse: *if your pet has four legs, it probably is a cat*. The reason for this is that "having four legs" is not specific to cats. Your pet could just as well be a dog or any other animal with four legs.

It would be different if the conditional probability *if your pet is not a cat, it probably does not have four legs* were true. In that case, your intuition would assume that *having four legs* is specific to cats. So if your pet has four legs it probably is a cat — the inverse of the original conditional probability! And since being a cat and having four legs seems almost equivalent, our intuition would then simply equate both probabilities. And this is exactly what happens in the case of the mammography test. Intuition treats a positive outcome as specific for having breast cancer (after all, if you don't have breast cancer, you will probably test negative) and therefore equates the two probabilities, regardless of the base rate.

The False-Positive Paradox

Accuracy Versus Reliability

Now that the solution to the mammography problem is within our grasp, even for those of us who consider themselves to be "bad at math", a second surprise awaits us. The probability of having breast cancer *if you test positive* is clearly not the same as the probability of testing positive *if you have breast cancer*. The latter is 80%, but the former is 80/1070, which is just 7.3%!

Contrary to our intuition, the accuracy of the mammography (80%) and the reliability of a positive outcome (7.3%) turn out to be two totally different things. Paradoxically, even a fairly accurate test can produce totally unreliable results. Of all the positive outcomes, no less than 92.7% are false positive. So what about the probability of you having the new disease? How reliable is the outcome of your test?

Are You Carrying the New Disease?

Assume your hometown has 10,000 inhabitants. Since 1% of the population is infected, about 100 out of the 10,000 inhabitants carry the new disease (in reality, of course, this number could also be 99 or 101). So 9,900 people in your hometown do not have the disease.

With an accuracy of 95%, the screening test will most likely pick out 95 of the 100 infected inhabitants. The test will also detect 95% of the non-infected. Only 5% of the people who don't carry the disease will test positive. But since a majority of 9,900 inhabitants do not carry the disease, even 5% results in a significant number of false positives: about $9900 \times 0.05 = 495$ people who are not infected will still test positive.

The reliability of the positive outcome of your test can now easily be calculated. It is the number of true positives (95) divided by the total number of positive outcomes (95 + 495). The answer is about 16%. So the complementary probability that you are not carrying the disease is not 5%, as you initially may have thought, but 84%! In other words, the result of your test is very likely a false positive.

You feel relieved, but at the same time you are surprised. How can a test that is 95% accurate yield such an overflow of false positives? The key to solving this apparent paradox is the *prevalence* of the disease or, as mathematicians say, its *base rate*.

Base Rate Matters

If the base rate of the disease is low — for instance, only 100 out of 10,000 people carry the disease — then the number of true positives will necessarily be relatively small. At the same time, since there are relatively many people who are not infected, many more than people who are infected, the number of false positives will be comparatively large. As a result, even an accurate test can produce more false positives than true positives. And this is what happened.

Now we can also understand why the doctors were initially perplexed by the results. After all, a total number of 590 out of the 10,000 inhabitants tested positive, far above the expected average of 1%. Could this be the start of an epidemic? Or had something gone wrong with the testing? After a restless night, the doctors came to realize what had gone wrong. The lower the base rate, the larger the proportion of non-infected participants, and therefore the higher the number of false positives. As a result, the majority of the 590 positives were false positives.

Covid-19

Test, Test, Test

April, 2020. The streets are deserted. Only once in a while an empty bus, still running on schedule, passes by. Out there, an invisible enemy is on the loose, holding everyone hostage in their homes. The evening news reports each day's tally of infections and deaths. Like wildfire, the virus has spread around the globe.

"We cannot fight a fire blindfolded", the General Director of the World Health Organization proclaims. "We have a simple message for all countries: *test, test, test*". Track and trace. Is it really as simple as it sounds? Apart from practical obstacles, the path may prove itself a dead end. As we've learned, even an accurate test will be useless if it generates too many false positives.

Two types of tests are currently available. PCR tests, also called swab tests, and antibody blood tests. Swab tests detect active infections by tracing viral RNA. Blood tests detect past infections by tracing antibodies in the blood.

With a swab test, you don't want false negatives. You don't want infected people who think they are not infected unknowingly spreading the disease. Conversely, with a blood test, you don't want false positives. After all, you don't want people on the streets who think they are immune, but actually are not. The results of the tests must therefore be as reliable as possible.

The Sense and Nonsense of Mass Screening

In order to get reliable results these tests must be accurate. The swab test should detect as many of the infected people as possible. Technically speaking, the *sensitivity* of the test must be high. The blood test, on the other hand, should exclude as many people as possible who do not carry antibodies. Technically speaking, the *specificity* of the test should be high.

Intuitively, you would think that an accurate test — say a test with a sensitivity of 96% and a specificity of 94% — would produce reliable results. After all, the margins of error (4% and 6% respectively) are quite small. Mass screening may therefore seem tempting to many decision makers. But, as we have seen, whether a test produces reliable results does not only depend on its accuracy; it also depends to a surprising extent on the base rate of the target.

What would happen if countries screened their populations in an attempt to identify and isolate infected people, as urged to by the head of the WHO? Clearly, such screenings only make sense when the epidemic is still manageable, i.e. when the prevalence of the disease is still low. But a low prevalence means that the swab test will lead to an excessive number of false positives.

If 10 out of 1,000 people are infected with the virus, a test as accurate as the above will result in 6 times as many false positives as true positives. And if the test were to be repeated on people who test positive, about 1 out of 4 positive results would still be a false positive.

It is the tragedy of large-scale screening that when the disease is still manageable medical tests are the least reliable.

The Immunity Passport

Unable to track and trace all people infected by the virus individually, countries around the world have enforced a lockdown, a nationwide quarantine to stop or at least slow the spread of the disease. People are being told to stay home. Schools and universities, bars and restaurants, cinemas, theaters, and museums — in short, all public places and recreational venues — are closed. Only vital businesses have been allowed to remain open.

The lockdown is proving effective in managing the disease, but economies have practically come to a standstill, unemployment is soaring, and social unrest is increasing. Calls to ease restrictions are becoming louder and louder. But if we were to lift all measures, we would run the risk of an outbreak that would be even harder to control.

But what about the idea of introducing an immunity passport? The restrictions could be lifted for those who have recovered from the disease and now have antibodies against the virus. They would get an "immunity passport", which would give them permission to work and travel. Thousands could go back to work, thousands could "regain their freedom". Who would not want the privileges that come with such a passport? A tremendous demand for antibody blood tests is to be expected.

Currently, the market is already flooded with antibody blood tests, competing in accuracy and ease of use. You can buy a Covid-19 home test kit and know the result within 24 hours. Politicians, travel agencies, and employers are all under the spell of immunity testing. But that spell is nothing more than a collective delusion. Simple math shows why.

Say an antibody test has a sensitivity of 94% and a specificity of 96%. If someone has antibodies to the Covid-19 virus, the test will be positive with a probability of 94%. If he or she has no antibodies, the test will be negative with a probability of 96%. Again, on the surface, the test may appear to yield reliable results. Now suppose that the test is widely available and a run on the test occurs. What would be the result?

It all depends on the relative number of people that have been infected in the past and have since recovered. In London, for instance, the proportion of the population with a past infection is estimated to be about 10%. So 100 out of 1,000 would carry antibodies to the virus. Of these, about 94 would be detected by the blood test. Of the 1000 – 100 = 900 people who do not carry antibodies, about 36 (4%) would still test positive.

So your chance of testing positive without actually being immune is 36/(36 + 94) = 0.28, that is 28%. About 1 out of 4 people tested would get a passport, even though they are not immune at all! In New York City, where about 20% of the population is known to have been previously infected, this would be about 1 out of 7 people. And these are populations with relatively high rates of people who have had a Covid-19 infection in the past. What if the proportion of the population with a past infection is only about 5%?

If 500 out of 10,000 people (5%) carry antibodies to Covid-19, about 470 people (94%) will test positive. Of the remaining 9,500 people who do not carry antibodies to Covid-19, 380 (4%) will also test positive. As a result, 44.7% would be false positive. Only if everyone who tests positive is tested a second time, and the probability of the outcome is independent of the previous result, would most false positives be filtered out.

Instead of using these tests as a tool for the large-scale diagnostic screening of individuals, we would be better off using them on large random samples. Once we know the accuracy of the tests, we can filter out the false positives and learn about the prevalence of the infection in the population. If these random screenings are carried out regularly, we will be able to monitor the disease and hopefully be better prepared for a potentially deadlier second wave.

4 Trials by Numbers

The Trial of the Century

On June 13, 1994, the bodies of Nicole Brown and her boyfriend, Ronald Goldman, both brutally stabbed to death, are discovered by a neighbor. When a bloodstained glove is found behind the house of Nicole's ex-husband, famous actor and former American football player O.J. Simpson, he is charged with double murder. On June 17, more than a thousand reporters await his arrival at the Los Angeles police station. But Simpson doesn't show up.

Simpson is tracked down riding in a white Ford Bronco on the I-5 freeway. The ensuing car chase involving twenty police cars and several news helicopters is covered on live television by ABC, NBC, CBS and CNN. Millions watch. Thousands of people on the streets of Beverly Hills witness Simpson's ultimate arrest.

On November 9, 1994, the case of the People of the State of California versus Orenthal James Simpson begins. The media attention is overwhelming. The glamor of a celebrity and the horror of the crime, spiced with police racism and O.J.'s history of spousal abuse, are more than enough to turn the televised trial into a national media spectacle. For eleven months, America is under the spell of what the media calls "the trial of the century".

The Trial

Although the murder weapon, a knife, is never found and there are no witnesses to the murder, Simpson's DNA is detected in several traces of blood from the crime scene. The prosecution is convinced that the DNA is strong evidence. After all, the probability of Nicole having been killed by someone else with almost the exact same DNA profile as her former partner is negligible.

But O.J. Simpson is supported by a team of top lawyers. They soon become known as the *Dream Team*. The team successfully undermine the reliability of the DNA evidence. They accuse the lab investigators of having mishandled the material, while they allege that the Los Angeles Police Department, driven by racist motives, tampered with the blood samples.

The district attorney, however, has another iron in the fire. Simpson has a history of spousal abuse. On multiple occasions, he has been investigated by the police after Nicole called 9-1-1. In court, the DA shows photos of her battered face. Simpson is portrayed as a woman abuser, unable to control his aggressive outbursts of jealousy. For the former football hero there is no line between physical violence and murder. "A slap is a prelude to homicide", as one prosecutor puts it. But the defense cleverly counters by appealing to the odds.

Every year, they respond, between three and four million women in America are battered or otherwise physically abused by their former or current husbands and boyfriends. According to the 1992 FBI Uniform Crime Reports, 913 women were killed by their husbands and 519 by their boyfriends. This means, the Dream Team contends, that less than 1 in 2,500 cases of spousal abuse results in murder. That's barely 0.04%! This figure convincingly refutes the prosecution's argument. Simpson's physical abuse of Nicole Brown was not a prelude to her murder. O.J.'s history of spousal abuse is clearly irrelevant.

The Verdict

On the morning of October 3, 1995, the trial, which has lasted eleven months, comes to an end. The LAPD is put on the highest state of alert. All over America people take a break from their usual daily activities. The phones stop ringing. On the New York Stock Exchange, trading drops 40%. About 100 million people settle in front of the television or near the radio, eager to witness the verdict of the century live. After only three hours of deliberation, the jury comes out. The verdict of the jury is unanimous. O.J. Simpson is *not guilty*.

Deceived by Numbers

Of course, no one knows what was discussed by the jury during those three hours. Jury deliberations are kept secret. But imagine that you were on the jury, discussing O.J. Simpson's history of abusing his ex-wife Nicole.

The defense seems to have convincingly countered the prosecution's claim that "a slap is a prelude to homicide". If spousal abuse were a prelude to murder, millions of women in the U.S. who are the sad victims of spousal abuse would be killed by their abusive partners. In fact, only a small minority of spousal abuse cases result in murder. The statistic of 1 in 2,500 proves that O.J.'s violent past is irrelevant. The prosecution's portrayal of a jealous and violent O.J. killing his ex-wife may sound plausible, but *numbers don't lie*.

But imagine now that instead of being a member of the jury, you are a critical journalist. You have been following the case closely. You know a thing or two about mathematics and you are aware that statistics themselves are never hard. They always ask for a correct and precise interpretation. So what does this 1 in 2,500 statistic actually tell you?

It tells you that every year about one in 2,500 men who batter their girlfriends or wives actually go on to kill them. At first glance, this seems to effectively disprove the prosecutor's line of argument. At the same time, murder and physical abuse seem somehow related. Then, all of a sudden, it becomes clear. The defense's response was the right answer to the wrong question!

The defense answers the question as to what proportion of all abused women are murdered by their abusing partner. But what you want to know is what proportion of all abused *and* murdered women are murdered by their abusing partner. In other words, what you want to know is more or less the inverse.

Nicole Brown *was* murdered. Therefore, the real question is whether she was murdered by O.J. Simpson or someone else. Clearly, the statistic presented by the defense is irrelevant. After all, it is not only rare for spousal abusers to kill their wives or ex-wives, but it is also rare for other people to kill the abused wives of others. We have to know how these statistics relate to each other. Let's do some journalistic research.

The True Numbers

According to the 1992 FBI Uniform Crime Reports, 913 women were killed by their husbands and 519 were killed by their boyfriends. So a total of 913 + 519 = 1432 women were killed either by their husbands or by their boyfriends. We know from the defense that this figure corresponds to approximately 1 in 2,500 abused women. That's 40 out of 100,000. Now let's see how many women were killed by someone other than their abusive partner.

The same 1992 FBI Uniform Crime Reports indicate that a total of 4,936 women were murdered in 1992. Since 1,432 of these women were murdered by their partner, 4936 – 1432 = 3504 women were murdered by someone other than their partner.

The total American population of women at that time was about 125 million. So 3,504 of 125 million women were murdered by someone other than their partner. That is about 2.8, say 3 out of 100,000.

Let's assume that all women, whether or not abused by their partner, are at the same risk of being murdered by someone other than their partner. Now it's easy to calculate the proportion of all abused *and* murdered women who were murdered by their abusive partner.

Out of 100,000 abused women, 40 women were killed by their abusive partner and 3 women were killed by someone else. So 40 abused women out of a total of 40 + 3 = 43 were murdered by their abusive partner. That is about 93%. In other words, not 1 in 2,500 but at least 9 in 10 times the abusive partner is the murderer!

As a journalist you realize that this statistic will make big headlines. After all, it implies that there is good reason to suspect the former American football star of having murdered his ex-wife. But does the statistic mean that Simpson is actually guilty? No, it doesn't. Even if the statistics are correct and, what's more, correctly applied, they cannot eliminate the need for tangible evidence, as will become clear in the next chapters.

Confusing Conditional Probabilities

Why was the jury so easily misled by the low statistic used by the defense? That's because it is far easier to grasp the probability of a possible outcome (murder) of a certain scenario (spousal abuse) than the probability of a possible scenario (spousal abuse) given an observed outcome (murder). Our intuition sees the former as a shortcut for the latter. And since only 1 in 2,500 cases of spousal abuse results in murder, our intuition dismisses this scenario outcome as highly improbable.

In fact, it was the prosecution itself that had prepared the ground for the defense by focusing on the scenario that "a slap is a prelude to homicide". In graphic detail, it presented Simpson's history of brutally abusing Nicole Brown to the jury in an attempt to show that Simpson was only a step away from murdering her. But, as one of the defense attorneys later wrote: "We knew we could prove, if we had to, that an infinitesimal percentage — certainly fewer than 1 out of 2,500 — of men who slap or beat their domestic partners go on to murder them".

This low percentage the defense came up with effectively beat the prosecution's one-liner. After the verdict, jurors called the prosecution's focus on domestic violence a "waste of time". The fact of the matter was that the jury confused two conditional probabilities: the probability that spousal abuse leads to the murder of the abused woman and the probability that the murder of an abused woman was part of a history of spousal abuse.

Epilogue

Twelve years after the "not guilty" verdict, a ghostwriter published a book based on interviews given by O.J. Simpson. The proceeds from the book went to Ron Goodman's family. The title of the book was *If I Did It*.

J'Accuse…!

Clearly, not only in the medical world, but also in court we encounter the confusion of conditional probabilities. What's more, legal practice has its own variant, a variant which is so notorious it has even been given a name of its own: the *prosecutor's fallacy*. The fallacy is so named because it is mostly committed, intentionally or not, by the prosecutor. In short, it goes as follows. If it is highly improbable that an innocent person is by chance involved in a particular incident, then the person involved must be guilty.

The prosecutor's fallacy is behind some of the greatest miscarriages of justice in the history of law. One of the earliest cases where probability theory is misused to convict an innocent person is the famous Dreyfus Affair. The conviction of the Jewish Captain Alfred Dreyfus for treason was, in the words of the great French novelist Emile Zola, "one of the greatest iniquities of the century".

The Prelude

Around the turn of the nineteenth century, nationalism and anti-Semitism were violent sentiments in France. In 1870, France had been defeated by its arch-enemy Germany. The eastern French region of Alsace, including its capital of Strasbourg, was annexed by the German Empire. An arms race with Germany ensued. Strong feelings of revenge dominated French society for decades.

At the same time anti-Semitic riots were not uncommon. Although Jews were highly integrated into French society, many nationalists believed that they were conspiring against the French Republic. Anti-Semitic newspapers such as *La Libre Parole* had a large circulation.

Such was the climate when in 1894 a French housekeeper, employed by the German Embassy in Paris but secretly working for the French military, found a torn-up anonymous memo in a wastepaper basket. The document only referred in general terms to other documents about a new piece of French artillery, yet the French Minister of War and the General

Staff saw an opportunity to improve their damaged public image. They went looking for a suitable scapegoat.

Soon an ideal individual had been found: Captain Alfred Dreyfus, a Jewish artillery officer of Alsatian origin and former employee at the General Staff. Without any evidence, Dreyfus was arrested to be court-martialed. *La Libre Parole* began writing about the affair. It was the start of a ruthless press campaign to influence public opinion before the trial. A secret file was concocted, evidence fabricated, and rumors spread. All the conspirators had to do was to link Dreyfus to the memo.

A Mathematical Proof

Several experts were called to testify that the memo was written in the handwriting of Alfred Dreyfus. By far the most authoritative expert witness was Alphonse Bertillon, the head of the Paris Police Bureau of Identification. Although he was not an expert on handwriting, he testified that the handwriting of the memo was very similar to Dreyfus' handwriting.

As if this were not enough, Bertillon claimed that Dreyfus had deliberately imitated his own handwriting, so that if he was caught he could say that it was not he who had written the memo, but someone else who had imitated his handwriting. And for this bizarre theory of auto-forgery Bertillon had mathematical proof.

In the text of the memo, Bertillon noticed, were 13 polysyllable words that appeared twice. When you laid a grid of vertical lines of a certain spacing on top of the memo, it turned out that four of these 13 pairs of polysyllable words either started or ended in exactly the same position relative to one of the gridlines.

Bertillon estimated the probability of such a coincidence in a natural handwritten text at 1/5. The probability that it happened four times was, according to Bertillon, simply $(1/5)^4 = 1/625$. This probability of only 0.16% was so small that the text couldn't be a natural handwritten text. It had to have been fabricated. And this, according to Bertillon, mathematically proved his hypothesis of self-forgery. Evidently, Dreyfus was the shrewd author of the memo!

Bertillon was, to say the least, not an expert in probability theory. That didn't matter. Although none of the judges at the closed court-martial seem to have understood his "mathematical proof", its scientific appearance gave the verdict a façade of objectivity.

Devil's Island

Dreyfus defended himself bravely with the support of a dozen defense witnesses, all in vain. Already convicted by public opinion, he was sentenced to life imprisonment and transferred to Devil's Island, a small uninhabited island off the coast of French Guiana.

He was the only prisoner on the island and was kept in a 4-by-4-meter hut, completely isolated from the outside world, except for some heavily censored letters from his wife Lucie. And if solitude, despair, and fevers — the island was infested by mosquitoes — did not make his life into a hell, the commander of his guards did. For four long years he imposed an inhuman regime on his prisoner, who was widely regarded as a traitor to his country.

J'Accuse...!

In the meantime, Alfred's older brother Mathieu never gave up the arduous fight for his brother's liberation. Over time, a number of prominent people joined the fight for a new trial. Then the real culprit was discovered by a French colonel. The perpetrator was tried but acquitted after just three minutes of deliberation. Instead, the colonel was arrested.

The world-famous novelist Émile Zola was outraged. He wrote an article in which he accused in detail and by name all those responsible for "one of the greatest iniquities of the century". The article appeared on January 18, 1898, on the front page of the newspaper *L'Aurore* and made history with its headline *J'Accuse...!* Zola had to flee to Great Britain to escape prison, but countless people now believed that Dreyfus was in fact innocent. The conspirators began to lose control of the situation.

In 1899 the Supreme Court overturned the judgement of 1894. Dreyfus, who was completely unaware of the extent of what had become known as “The Affair”, was transported back to France and had to stand trial again, this time at the military court in Rennes. He was convicted of treason and sentenced to ten years’ imprisonment, but received a presidential pardon on the condition that he accepted the guilty verdict. Dreyfus accepted and was finally released. A law was filed granting amnesty to all those involved in the affair.

A Commission of Mathematicians

The affair seemed to be closed for good. However, when in 1902 the elections strengthened the position of the progressive parties in parliament, the Dreyfus Affair once again became the subject of investigation. Three mathematicians from the prestigious *Académie des Sciences* were commissioned by the Supreme Court to write a report on Bertillon’s so-called mathematical proof. The commission was headed by the most prominent French scientist at that time, Henri Poincaré.

Henri Poincaré (1854–1912) was a well-known public figure in France because of his international reputation and his popular books on the interpretation of science. For ten years he had held the chair of probability calculus at the Sorbonne. He was regarded as the top specialist in the mathematics of randomness in France.

Poincaré and his two colleagues concluded that Bertillon’s system was devoid of any scientific value. In the report Poincaré writes that it is hard to explain to anyone who was not aware of the history of the affair the need for a lengthy discussion about a system that is so evidently absurd. In addition to the erroneous reconstruction of the torn-up memo, the rules of probability were not correctly applied, nor was their application to the case legitimate.

The Wrong Number

Even the most superficial examination reveals that the small probability Bertillon came up with is wrong. Instead of the probability of 4 coincidences in 26 cases, that is, the total number of beginnings and endings of the 13 pairs of polysyllables, Bertillon had calculated the probability of 4 coincidence in 4 cases.

Suppose, for example, you throw 4 sixes when rolling a die 26 times. Then the elementary mistake Bertillon made is calculating the probability of this to be $(1/6)^4$. But that would be the probability of throwing 4 sixes in 4 rolls. The probability of throwing 4 sixes in 26 rolls is of course much larger. And the same applies to the probability calculated by Bertillon. The correct probability, Poincaré notes, is about 400 times greater.

But even if we grant Bertillon the small probability he came up with, it is still not legitimate to conclude that this probability is so small that the coincidence could not have happened by chance. Think of a lottery. The probability beforehand that the number 25 will show up may be small, but that does not imply that the draw isn't fair.

Instead of focusing on the one coincidence that he observed, Bertillon should have widened his view to encompass all possible coincidences in the memo that he would have found equally remarkable had they occurred.

Suppose, Poincaré writes, you apply a coordinate system to the memo and define all letters in the memo according to their coordinates. Then the probability of finding coincidences that are equally remarkable to "a spirit as attentive as Mr. Bertillon" is almost certain.

The Wrong Method

But not only did Bertillon come up with a wrong number, the method he used was also completely wrong. Nowadays, we would say that Bertillon had committed the prosecutor's fallacy. He stated that the probability of the coincidences happening by chance (implying that Dreyfus is innocent) was so small (1/625) that one could safely conclude that the memo had been fabricated (implying that Dreyfus was guilty).

Now, calculating what the probability is of an observed coincidence in a certain scenario is fundamentally different from calculating what the probability is that a certain scenario led to an observed coincidence. Poincaré illustrates the difference using the example of drawing balls from an urn.

The probability of randomly drawing a white ball from an urn with 90 white balls and 10 black balls is easy to calculate. It's 90%. But calculating the probability that a white ball was drawn from one specific urn out of several urns, each with its own ratio of white and black balls, is not so easy! You need to know for each specific urn what the probability is of drawing a ball from that urn. You also need to know the ratio of balls in each urn.

In the same way, we need to know the different scenarios that can lead to the observed coincidences in the memo. And of each of these scenarios, we need to know its probability as well as the probability that it will lead to the observed coincidences.

The Danger and Futility of a Trial by Numbers

Poincaré saves his most fundamental blow to Bertillon's "mathematical proof" for last: even if Bertillon had applied a mathematically correct method, it would have been illegitimate. The reason is that the scenarios we face in court — as, for instance, the different ways of forging the memo — are not chance events but what Poincaré calls *moral elements*. By this he means events that belong to the domain of human behavior and therefore escape any form of mathematical calculation.

It is dangerous and pointless to replace such moral events with numbers or, as we argued in Part One, to quantify the plausibility of scenarios in terms of mathematical probability. Probability calculus, writes Poincaré, is no substitute for having common sense and good judgement. If common sense had prevailed, the judges should have observed that, if the author of the memo had really wanted us to believe that the document was a forgery, he would no doubt have chosen a system that could be detected by the experts more easily, not Bertillon's convoluted system!

Rehabilitation

On July 12, 1906, the Supreme Court unanimously annulled the earlier judgements of the military courts. The Court judged Bertillon harshly. In its words, the head of the Bureau of Identification had "reasoned badly on forged documents". His mathematical proof was a meaningless intellectual construction.

The Supreme Court ruled that Captain Dreyfus had to be fully rehabilitated. The next day Dreyfus was reinstated in the army and promoted to Major. He served his country in World War I and was made Officer of the Legion of Honor. He died on July 12, 1935 at the age of 75.

An Angel of Death

The Dutch Dreyfus Affair

Mathematics was used to "prove" that Dreyfus was guilty. The truth is, it was bad math that was abused to commit the prosecutor's fallacy. In its usual form the prosecutor's fallacy states that "the probability of X happening if a person is innocent is so small that we can safely rule out that the accused is innocent." What happens here is that the probability that someone is innocent given the incident is equated with the inverse probability that the incident occurs given that someone is innocent. And this probability is typically very small.

The prosecutor's fallacy is itself far from innocent. It underlies some of the severest miscarriages of justice ever. In the United Kingdom, the trial of Sally Clarke, which took place in the 1990s, is a well-known example. Her two children both died a tragic sudden infant death. This was considered so unlikely that Sally had to stand trial as the murderess of both her children and was sentenced to life imprisonment.

At the beginning of this century, a miscarriage of justice similar to the Sally Clark case took place in the Netherlands. During the shifts worked by the pediatric nurse Lucia de Berk an apparently high number of unexplained deaths among patients occurred. Although it was a tragic coincidence, Lucia was depicted in the media as a horrific serial killer of innocent children and elderly people, and was eventually sentenced to life imprisonment. The case is sometimes called the "Dutch Dreyfus Affair" because of its similarities.

A Killer Nurse

On September 11, 2001, the very same day Al Qaeda launched its deadly terrorist attacks on the United States, a remarkable press conference was held in the Juliana Children's Hospital in the Dutch city of The Hague. The conference was broadcast by the regional radio station. The management of the hospital had a shocking announcement to make. One of the pediatric nurses was involved in several suspicious deaths and near-deaths of patients at the hospital. The hospital apologized and expressed its sympathy to the parents of the deceased children.

The Dutch newspaper *De Telegraaf* almost immediately published the horrifying story of a nurse who had allegedly killed a large number of young patients. In the same article, the director of the hospital conveyed his condolences to the families of the young victims and stressed that he would do everything he could to ensure that justice was done. Other newspapers followed. The general public became increasingly convinced that a Dutch pediatric nurse actually was a horrific serial killer.

"One in Seven Billion"

An investigation that was carried out by the hospital itself revealed that in the period during which the nurse was employed at the hospital, eight unexplained deaths and near-deaths had taken place, all of them occurring during her shifts. Not a single case had occurred when the nurse was not on duty. In addition, the investigation revealed that in the same period a number of unexplained deaths had also occurred during the nurse's shifts on two wards for elderly patients in the Red Cross Hospital where she had been employed as well.

The hospital estimated the probability of its being pure coincidence that the nurse had been present at all these "incidents" at only 1 in 7 billion. Based solely on this extremely small probability — there was no medical evidence — a police investigation was launched. Soon the "1 in 7 billion" probability was leaked to the media, where it was interpreted as proof that the nurse was guilty. To the general public she became known as *The Angel of Death*.

A Witch Trial

On December 13, 2001, 40-year-old Lucia de Berk was arrested while sitting at her grandfather's deathbed. A few months later she was tried. She was portrayed by the prosecution as an immoral fraud. Her short period of work as a call girl in Canada at the age of 17 and her forgery of a Canadian school diploma qualifying her to be admitted to the Dutch nursing school were presented as moral "evidence". Ultimately, she was deemed capable of unspeakable crimes. In the court drawings that appeared in the newspapers, the gentle woman was invariably depicted as a witch.

These imputations, of course, were not evidence in themselves. Since there had been no witnesses and no traces of poisoning or suffocation, it was crucial to prove that her presence at the large number of unexplained deaths and near-deaths could not be a coincidence. An expert was required to provide a precise number, underpinned by mathematical calculations. A professor of law with a degree in statistics was called to court.

Mathematical Proof

Assuming that the shifts and deaths were randomly distributed, the law professor used simple statistics to calculate the probability that the relevant deaths and near-deaths would coincide with Lucia's shifts.

For the incidents that occurred at Juliana Children's Hospital, he calculated that the probability was less than 1 in about 9 million. Taking into account all 27 nurses, he corrected this number to 1 in about 350,000. For the two other wards he arrived at probabilities of about 1 in 14 and about 1 in 73, respectively. Next, the law professor multiplied the three values and arrived at a probability of approximately 1 in 342 million.

All Dutch newspapers and many European newspapers went to press with this 1 in 342 million probability. To the general public it was clear evidence that Lucia was guilty of the unprecedented crimes. Lucia herself testified that, in spite of the incredibly small number, she still believed that her presence had been a matter of coincidence. In a reaction, the law professor addressed the judge with the words: "Honorable Court, it was not coincidence. The rest is up to you".

Life Imprisonment

On March 24, 2003, Lucia was sentenced to life imprisonment. She was held responsible for the murder of four patients and the attempted murder of three more. Assuming that it was no coincidence that the deaths had occurred during Lucia's shifts, the judge found her guilty in all cases where there was any evidence and in all cases that, according to a medical expert, could not be explained by natural causes.

A year later, in the Court of Appeal, Lucia was convicted of seven murders and three attempted murders. The prosecution had come up with new evidence. One of the children was said to have been poisoned with digoxin. Lucia would have had the "means and opportunity" to kill the child. Another very sick child had died after being prescribed a fairly high dose of chloral hydrate. Again, Lucia would have had the "means and opportunity" to administer an overdose.

The court considered the two cases of alleged poisoning by Lucia to be proven. The other eight cases were deemed to be proven by what is referred to as *chain-link proof*. Since two murders were deemed to have been proven beyond reasonable doubt, weaker than normal evidence was sufficient to prove that the other cases were also murders or attempted murders. Lucia was sentenced not only to life imprisonment, but also to compulsory psychiatric treatment, in spite of the fact that no evidence of mental illness had been found.

The Chain-Link Proof

The Court of Appeal explicitly stated that "statistical evidence" played no role in Lucia de Berk's conviction. However, as the Dutch physicist and Nobel Prize winner Gerard 't Hooft pointed out, if this had been the case, the chain-link proof would lack any grounds.

The chain-link proof presupposed that the other (near-)deaths shared a common feature with the ones that were considered proven: the so-called link. But there was no medical evidence that the other "suspicious" incidents were also murders or attempted murders. They were only labelled suspicious because of Lucia's presence. The only link was the "mathematical proof" that her presence was no coincidence!

All Hope Forlorn

Finally, an appeal was made to the Supreme Court of the Netherlands. Important new evidence was presented by the defense. A report from a state-of-the-art forensic laboratory in Strasbourg showed that the digoxin poisoning was based on the erroneous identification of a very similar but harmless chemical.

The Supreme Court, however, only ruled on the correctness of the procedure followed by the Court of Appeal. The Court judged that the combination of the two sentences, life imprisonment and compulsory psychiatric treatment, was incorrect and sent the case back to the Court of Appeal to be re-evaluated on the same grounds as before.

After the Supreme Court's ruling, Lucia suffered a stroke. The prison guards did not take her condition seriously and it wasn't until ten hours later that she was transferred to the prison hospital. She had lost the power of speech and all motion on the right side of her body when, on July 13, 2006, the Court of Appeal delivered the definitive verdict. Lucia was sentenced to life imprisonment without parole.

The Fight for Justice

Lucia was physically and mentally broken. Despised throughout the country, she would have lost any prospect of freedom if it had not been for the sister-in-law of the head pediatrician at Juliana Children's Hospital. Herself a geriatric doctor, she had become interested in the case and had begun examining the medical records in 2004. To her surprise, not one of them suggested murder. Together with other family members, she started a fight for Lucia's release.

The fight gained momentum when her brother, a prominent professor of philosophy, joined her committee. He wrote a book exposing many of the wrongs of the case down to the smallest detail. He observed that Lucia had never had the opportunity to poison the child with digoxin.

Moreover, the autopsy had not shown any of the usual signs of digoxin poisoning. Along with the findings of the forensic lab in Strasbourg, this clearly indicated that the child had not been poisoned with digoxin at all. Nor had the other child received an overdose.

These findings nullified the main reason for Lucia's conviction. They clearly also undermined the chain-link argument. A request for the reconsideration of Lucia's conviction was submitted to the Committee for the Evaluation of Closed Criminal Cases. The Committee issued a report recommending that the case be reopened. A petition for Lucia's release appeared as a full-page newspaper advertisement. It was signed by 1,300 people, including many important scientists and scholars from all over the world.

Freedom and Rehabilitation

Six months later, Lucia was granted a temporary suspension of sentence. After more than six years in prison and still physically impaired, Lucia was free at last. Two years after her release, on April 14, 2010, the Public Prosecution capitulated and the *not guilty* verdict was delivered.

The life of an innocent woman was destroyed. Justice and truth had been trampled underfoot. Both the prestige of the Dutch legal system and the reputation of the medical profession had been severely damaged. The Dutch Minister of Justice wrote Lucia a letter of apology. He described the course of events as horrific. From the Public Prosecutor's Office Lucia received undisclosed compensation. The hospital paid her 45,000 euros.

The Lucia de Berk case was one of the greatest miscarriages of justice in modern Dutch history. It was "the Dutch Dreyfus Affair". Perhaps its most conspicuous similarity was the misuse of mathematics to establish the guilt of an innocent individual.

"The Mathematical Proof"

Mathematics had played a crucial role from the very beginning. The police investigation was launched because of the "1 in 7 billion chance" that Lucia's presence had been a mere coincidence. The "1 in 342 million" figure calculated by the so-called expert witness circulated widely as clear proof of Lucia's guilt. As a Dutch professor of Criminal Law put it: "Statistical evidence has been of enormous importance. I do not see how one could have come to a conviction without it".

Imagine that the great mathematician Henri Poincaré has descended from the higher spheres of mathematics, where his soul undoubtedly lives on, and that he has been asked to provide a report on the calculations of the "Dutch Bertillon". Would this report look very different? Probably not. The data were wrong, the method was wrong, the outcome was wrong. The whole mathematical exercise was meaningless.

The Circle of Suspicion

Just as the reconstructed memo that incriminated Alfred Dreyfus should have accurately reproduced the original, so should the data used by the Dutch expert witness have been representative. A professional statistician would have ascertained that his data were neither biased nor selective. In fact, they were both biased and selective.

The data related to the patients at Juliana Children's Hospital were provided by the hospital itself. The hospital had examined Lucia's shifts and marked any unexplained deaths during one of her shifts as suspicious. None of these deaths had initially been flagged as suspicious. They had all been considered natural deaths at the time. In short, it was Lucia's presence alone that led to the deaths being retroactively deemed suspicious. These "suspicious (near-)deaths" were then used by the expert witness to test whether Lucia's presence was suspicious.

At the very least, it should have been investigated whether there were any unexplained deaths that should also have been classified as suspicious, but at which Lucia had not been present. And indeed, there were two other unexplained deaths which were as "suspicious" as those which had occurred during Lucia's shifts.

Does this matter? It does. The probability of Lucia being present 8 times out of 10 is much greater than the probability of her being present 8 times out of 8. Using the same method, the professor of law would not have arrived at the original 1 in about 9 million but at 1 in about 250,000. And since there are about 250,000 nurses employed in the Dutch Health

Care System, one would expect this tragic coincidence to happen once over time. The law professor's calculations did little more than provide the hospital's flawed investigation with a semblance of objectivity.

Mathematical Magic

Let's assume that the data had been correct and accurate, as the hospital itself claimed. Even then, the professor of law was completely off-base in applying a mathematical model that was far too simplistic.

The model he used can easily be envisioned as taking marbles from a vase or an urn. The marbles represent the nurses' shifts. Say the black ones are Lucia's shifts, and the white ones are the shifts of others. Each draw stands for a suspicious incident (a death or a near-death). Now, the probability that the eight suspicious incidents coincided with Lucia's shifts corresponds to the probability that eight marbles are drawn randomly and without replacement, and that all these eight marbles are black.

The probability of this happening is less than 1 in 9 million. The law professor multiplied this number by 27 since, if it were a coincidence, it could have happened to any of the 27 nurses employed on the ward. In this way he arrived at a probability of 1 in 350,000.

Next, he applied the same model to the two wards for elderly patients at the Red Cross Hospital where Lucia had worked in the same period. The outcomes he found were, as we saw before, about 1 in 14 and about 1 in 73, respectively. Not very spectacular.

Then the law professor performed some mathematical magic. He multiplied the three values he had found and thereby arrived at an incredibly small probability of 1 in 342 million. Clearly, he concluded, it could not have been coincidence.

The Absurdities of a Flawed Model

Can one legitimately multiply these different probabilities? The answer is no. In reality, one would expect that it would make little difference whether the suspicious incidents and shifts were spread over three wards, or whether they were all concentrated in one large ward. But if you use the same urn model and calculate the probability for the same suspicious incidents and shifts, but now merged into one ward, you arrive at a probability of 1 in 3.8 million, a number that is almost a hundred times larger.

So if Lucia had worked the same number of shifts on one ward instead of three, according to the professor's model it would have been a hundred times less likely that she was a serial killer. This number may also seem extremely small, but it is in the same order of magnitude as winning a lottery. A very tragic lottery in this case.

Let's put the tragic coincidence in perspective. The probability that a nurse does not suffer the same tragic fate as Lucia did is the complementary probability 1 – (1/3,800,000). There are approximately 250,000 nurses employed in the Dutch Health Care System. The probability that none of them would suffer this tragic fate would be $(1 - (1/3{,}800{,}000))^{250{,}000}$. That's about 93.6%. So the complementary probability that at least one of them would suffer this tragic fate would be 6.4%. A small but significant probability, and presumably a lot larger than the probability that any of them would turn out to be a serial killer!

Let's examine another absurdity of this simplistic urn model. Suppose that the rate of incidents is lower than normal. Then the probability that Lucia's presence was just coincidence is of course high. After all, if Lucia really was a serial killer, this would show in the rate of incidents.

But the model of the "expert witness" completely fails to take this into account. It only "answers" the question of what the probability is that, *if* there are a number of n "incidents", m or more of these n incidents coincide with one of Lucia's shifts. The model comes up with a very small probability. But in fact, as we will see, in the period Lucia had been working the incidence rate (n) *was* lower than normal! So we end up with a blatant contradiction.

Clearly, the law professor's urn model is flawed and far too simplistic. Simple methods give exact results. But reality isn't that simple. The more exact a number is, the more misleading it usually is.

Alternative Scenarios Dismissed

The figure of 1 in 342 million renders it highly improbable that Lucia's presence was coincidence. The number of nurses on earth is much smaller. The law professor himself remarked to the judge that it was no coincidence. So what was it? The professor named five other possible scenarios besides murder:

"Perhaps Lucia was an incompetent nurse …".

"Perhaps she used to get the patients that were most sick …".

"Perhaps she was assigned to special shifts (for example night shifts, as the majority of patients die at night) …".

"Perhaps someone else was also present during all the suspicious incidents …".

"Perhaps she was being framed …".

Yet how can one mathematically weigh up the probabilities of these scenarios? It's not possible. Nor can one mathematically assign a probability to the murder scenario. These are all events that escape mathematical calculation. And that's why the 1 in 342 million figure is meaningless.

What's more, assigning an "exact" number to the coincidence scenario and simply saying that the weighing of the other scenarios is "up to the judge" is like handing a loaded gun to a child. All the other scenarios were either denied by Lucia or considered too improbable. The murder scenario, no less implausible, had been adopted by the media and public opinion almost from the very beginning. The incredibly small (and incredibly exact) probability assigned to the "coincidence scenario" paved the way for the prosecutor's fallacy: Lucia had to be guilty.

Justice Blinded by Numbers

Henri Poincaré would no doubt have warned the judges not to be blinded by a mathematical number. When moral elements are involved, they should use their good judgement. Good judgement, however, was hard to find.

During the second trial, suspicious incidents were added to or removed from the list, depending on whether or not Lucia had been present. Unproven incidents were chain-linked. And, on top of all this,

Lucia was labeled as a dangerous mentally ill person. Poincaré would no doubt have been shocked by this lack of good sense.

The judges should have realized that on the hospital wards where Lucia had worked people unfortunately die in greater numbers than average, and therefore unexplained deaths occur there in higher numbers. All the unexplained deaths that occurred when Lucia was present were regarded as natural deaths at the time. No one identified them as suspicious.

The judges should have realized that if Lucia had really operated in such a slick way, leaving no trace and never being caught in the act, it is hard to imagine why she had been so careless as to commit all her crimes precisely during her shifts, linking the "suspicious incidents" to herself. A far more plausible explanation is that the "victims" were not murdered in the first place.

The judges should have realized that Lucia's presence at many of the unexplained deaths could actually very well have been a coincidence. Coincidences tend to cluster. Just like the first raindrops that fall on the ground. Several nurses claimed to have experienced situations that may have been less extreme, but were very similar to what happened to Lucia.

Every single one of these observations should have been a red flag. But what should really have opened the judges' eyes was that the number of unexplained deaths on the ward at Juliana Children's Hospital was about the same, yes, even higher in the period that preceded that in which Lucia worked her shifts. How can this be if there really was a serial killer at work?

The Prosecutor's Fallacy

How could it be that the judges — and not only the judges, but half the country — were deeply convinced that an innocent nurse was in fact a horrific serial killer? One reason is the way she was publicly depicted. Both in the media and in court she was demonized and stereotyped as a witch. It was not Lucia, but an "angel of death" who stood trial.

The other reason is the intuitive force of the prosecutor's fallacy. Either the nurse was innocent or she was guilty. If she was innocent, it must have been coincidence. Other possible scenarios are unexamined.

The intricacies of finding and evaluating alternative explanations belong to the domain of "slow thinking".

Typically, the probability of the coincidence scenario is very small. After all, if it weren't, it would possibly be a coincidence and there wouldn't be a case! Our intuition concludes that it is safe to rule out such a small probability. It is simply "too amazing to be a coincidence!" So it follows that Lucia can't be innocent. And since she is either innocent or guilty, she must have committed the horrific crimes.

Now this shortcut of our intuition may feel very compelling, but its neglect of alternative scenarios is dangerous and its logic is flawed. As we have seen before, improbable coincidences do happen. As the law of truly large number explains, they happen all the time. How misleading the prosecutor's fallacy can be, is illustrated by the following example.

DNA matches can be very tricky in the absence of other evidence. Imagine that a crime has been committed and traces of the perpetrator's DNA have been found on the victim. When the police search their DNA database a DNA match is found. The person in question is arrested. The likelihood of a DNA match is only 1 in 100,000. So the chance that someone innocent has a match is only 0.001%.

Can we conclude that this person is therefore guilty of the crime? Not if there is no other connection or further evidence. In a city of 1,000,000 people you expect there to be 10 matches. So while the probability that someone innocent has a match may be as small as 0.001%, when there is no other link to the crime the inverse probability is 9 in 10, that is 90%.

The lesson is that in court our intuition is not only a poor guide, but a dangerous one as well. One should beware of the prosecutor's fallacy and never substitute numbers for good judgement. Numbers *do* lie.

5 The Monty Hall Dilemma

Let's Make a Deal!

Monty Hall

It is February 18, 2013. Today is the 50th anniversary of the game show *Let's Make a Deal*. On this special occasion, the show will be hosted once more by the legendary Monty Hall. Monty Hall (1921–2017) was the host of the show from the 1960s until the beginning of the 1990s. The program was an enormous hit soon after it first aired. Monty Hall became a national celebrity. People didn't say, "Today *Let's Make a Deal* is on television," but "Today *Monty Hall* is on television."

An episode of Monty's show typically started with some minor deals between the contestants and the host. With each deal, a modest prize could be won. The show ended with the famous *Big Deal*, the very last game everyone was waiting for.

The Big Deal

On stage there are three identical doors, all closed. Behind one of the three doors awaits the show's grand prize: a shiny new car or an all-expenses-paid, first-class vacation, among other things. Behind the two other doors there is either nothing at all or sometimes a small or bizarre consolation prize, such as a goat.

The contestant chooses one of the three doors and takes up a position in front of this door. All the time, the host continues to increase the tension. *Is the contestant lucky today? Has he or she chosen the winning door?* Nobody knows, except for the host himself and his personal assistant.

The game now reaches its climax. The host has the assistant open one of the two other doors, always a losing door. The contestant is aware of this. The contestant is offered a final choice: either to stick with the same door, or to switch to the other remaining door. The contestant usually wavers for some time and then makes his or her final choice. The host has his assistant open the other remaining door and …

A Haunting Dream

Imagine we're back in 2013. It is Monday evening, February 18, and the anniversary episode of the game show begins. You're a huge fan of *Let's Make a Deal.* You watch every daily episode. That evening, Monty hosts the show once again. From the start, you've been glued to your television. When the Big Deal arrives, you almost feel like you're the contestant yourself. What would you do? Switch or stay?

The very same night, you have a dream. You have chosen one of the three doors. Monty's assistant opens one of the other doors. Now what will you do? Will you switch doors or will you stick with your initial choice? You keep wavering. People start yelling. *Switch! Stay!* The shouting gets louder and louder. You awake. It's your alarm clock. The dream vanishes in the light of the early morning. The dilemma, however, keeps haunting you all day.

Ask Marilyn

Marilyn vos Savant, who was once mentioned in the Guinness Book of Records because of her exceptionally high I.Q., writes a weekly column in *Parade* magazine. Her column *Ask Marilyn* is read by millions of people. In her column on Sunday, September 9, 1990, Marilyn, still ignorant of the storm that her answer will unleash, replies to a reader's question about a game show similar to *Let's Make a Deal*. The reader poses the following problem.

Suppose you're given the choice of three identical doors. Behind one door there's a car, while behind each of the other two doors is a goat. You pick one door and after one of the other two doors — always a door hiding a goat — has been opened, you're asked to choose again. What do you think, is it to your advantage to switch your choice of doors?

Marilyn's surprising answer is that it is. If you switch doors, your chance of winning the grand prize doubles!

A Surge of Disbelief

Marilyn's answer leaves many readers totally perplexed and provokes fierce discussions throughout the nation. Fans of Monty's game show, readers of Marilyn's column, and numerous mathematicians climb into the ring to join the battle.

In July 1991, the *New York Times*, devoting a front-page article to the controversy, reports that Marilyn's answer has been "debated in the halls of the Central Intelligence Agency and the barracks of fighter pilots in the Persian Gulf. It even has been analyzed by mathematicians from MIT and computer programmers from the Los Alamos National Laboratory in New Mexico".

Marilyn receives over ten thousand letters from readers of her column. These letters are not expressions of support. The vast majority — nine out of ten readers — insist that she is absolutely wrong. It doesn't matter whether you change doors or not. After all, with only two closed doors left, the odds are obviously fifty-fifty.

Remarkably, many of her critics are mathematicians, and not the least reputable either. A professor of mathematics, whose name we will not mention, writes: "[...] you blew it! Let me explain. If one door is shown to be a loser, that information changes the probability of either remaining choice, neither of which has a reason to be more likely, to 1/2. As a professional mathematician, I'm very concerned with the general public's lack of mathematical skills. Please help by confessing your error and in the future being more careful".

The professor — a crusader against mathematical illiteracy — is convinced he's right. And although the tone of his reaction is rather denigrating, it is still mild compared to the other criticism received by Marilyn from many of his colleagues.

One caring mathematician offers Marilyn consolation: "You made a mistake, but look at the positive side. If all those Ph.D.'s were wrong, the country would be in some very serious trouble". But Marilyn vos Savant made no mistake and, yes, all those Ph.D.'s were wrong ...

Switching Doors Matters

Intuitively, you may well side with Marilyn's critics. Once a losing door is opened, two doors remain. Behind one door is the car, and behind the other door is a goat. Clearly, the probability of the car being behind either door is one in two, like the flipping of a coin. So there is no reason for the contestant to switch to the other door.

This all sounds very convincing. But not much later Marilyn receives another letter from the same professor. The professor apologizes. In all modesty he admits that he was completely wrong. If you switch doors, your chance of winning the car doubles. How can this be?

The problem of the three doors, for obvious reasons, is known as the *Monty Hall dilemma*. In her columns Marilyn offers her readers three ways of solving the dilemma. We will also present three solutions that are not too complicated.

Before we do so, for the sake of clarity, we will mention some important assumptions. First, the car is placed at random behind one of three identical doors and only the host and his assistant know which door it is behind. Second, after the contestant has chosen a door, the assistant opens one of the other two doors, always one concealing a goat and never the one which is hiding the car. Third, the contestant is aware of this.

A Tree Diagram

Let's return to your dream and imagine you're the contestant. You pick one of the three doors. We will call this door *door A* and the other two doors *door B* and *door C*. Now there are three scenarios, each with an equal probability of 1/3. In scenario 1, the car is behind door A, the door of your choice. In scenario 2 the car is behind door B and in scenario 3 the car is behind door C.

The assistant now opens one of the other two doors, always one that has a goat behind it. In scenario 1 the assistant opens either door B or door C. In scenario 2 the assistant opens door C and in scenario 3 the assistant opens door B. Next, the host asks you whether you want to switch doors. What would you do? Should you stick with your initial choice, door A, or should you switch to the other remaining door?

Suppose you stick with door A. Then, in two of the three scenarios — scenarios 2 and 3 — you will win a goat. After all, in each of these two scenarios, there is a goat hiding behind door A. In other words, your chance of winning the car is 1 – (2 × 1/3) = 1/3 (33.3%), the same as it was before.

If you switch doors, in two of the three scenarios — again scenarios 2 and 3 — you will win the car. In other words, your chance of winning the car is now 2 × 1/3 = 2/3 (66.7%). So by switching doors you double your chance of winning the car.

A tree diagram may clarify this. In the tree diagram the three doors are arranged in such a way that the first door is door A, the door of your choice. Next to door A are door B and door C. The branches represent the possible scenarios. As you can see, switching doors increases your chance of winning the car from 1/3 to 2/3.

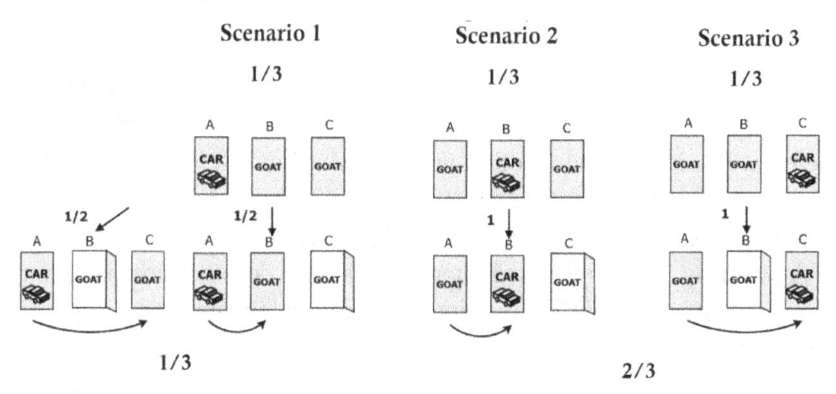

Using Natural Frequencies

An even easier way to approach the Monty Hall dilemma is the following thought experiment. Imagine again that you are the contestant. You pick one of the three doors. Your chance is one in three (1/3) that the car is hidden behind your door and two in three (2/3) that it's a goat.

What if you could repeat picking a door fifteen thousand times? Then, according to the law of large numbers, you would on average in 1/3 × 15,000 = 5000 trials pick a door hiding the car and in about 2/3 × 15,000 = 10,000 trials pick a door hiding a goat. In other words, if you don't switch, you will win the car on average only in 5000 of the 15,000 trials, but if you switch doors, you will win the car on average in 10,000 of the 15,000 trials. So when you're asked whether you want to stick with your choice or want to switch doors, you'd better switch doors!

But even after hearing these solutions, not everyone is convinced of this remarkable outcome. One of these disbelievers was the world famous — at least among mathematicians — Hungarian mathematician Paul Erdős (1913–96).

Paul Erdős and Computer Simulation

Erdős did not own a house of his own, but instead travelled throughout his life around the world, staying with mathematical friends at each stop along the way. During his stays, he and his hosts solved the most difficult mathematical problems. As a result, Erdős has a record number of more than 1,500 articles to his name, in all areas of mathematics. He ranks as one of the most prolific mathematicians of all time.

On a visit to his compatriot and colleague Andrew Vázsonyi (1913–2003) in 1995, the eccentric mathematician heard about the Monty Hall dilemma. His immediate reaction was that Marilyn vos Savant's solution could not possibly be true. According to the mathematical genius, changing doors should not make any difference.

A proof by Vázsonyi based on a tree diagram does not convince him. Annoyed, he asked his compatriot to give him an obvious reason for switching doors. Vázsonyi, not knowing how to convince Erdős by mathematical proof, resorted to a computer to simulate the game show.

He programmed the computer to determine at random behind which of the three doors the car had been placed and also which of the three doors was chosen by the contestant. He chose switching doors to be the contestant's strategy and had the computer run this simulation of the game no fewer than 100,000 times.

The law of large numbers guaranteed that the relative frequencies of the outcomes — either winning the car or winning a goat — would approach the underlying probabilities. The computer simulation confirmed that Marilyn's answer was correct. Erdős was finally convinced, although some whisper that he continued to ponder the matter of switching doors until his last breath.

The Turning Point

When Marilyn vos Savant publishes a second column about the Monty Hall dilemma, she provides a proof by means of a table with six possible outcomes. The reactions, however, are even more vicious and sarcastic:

> "… I am sure you will receive many letters on this topic from high school and college students. Perhaps you should keep a few addresses for help with future columns".
>
> "How many irate mathematicians are needed to get you to change your mind?"
>
> "May I suggest that you obtain and refer to a standard textbook on probability before you try to answer a question of this type again?"

Marilyn changes strategy. In a third column on the dilemma, she takes the same route Vázsonyi later did. She calls on students all over the United States to simulate the game by means of three paper cups and a penny, and to repeat each of the two strategies — switching doors and sticking to your choice — two hundred times. No fewer than 50,000 students answer her call. Almost all simulations result in one and the same outcome:

> "Our class, with unbridled enthusiasm, is proud to announce that our data support your position …".
>
> "At first I thought you were crazy, but then my computer teacher encouraged us to write a program, which was quite a challenge. I thought it was impossible, but you were right!"
>
> "I also thought you were wrong, so I did your experiment, and you were exactly correct. (I used three cups to represent the three doors, but instead of a penny, I chose an aspirin tablet because I thought I might need to take it after my experiment.)"

Marilyn is delighted. Of those who carried out the experiment with three paper cups and a penny, almost 100% confirm the correctness of her solution. Of those who wrote a computer program that simulates the game, almost 97% are now convinced that it pays to switch doors.

A lot of people who have not conducted the experiment now also write Marilyn to let her know that they have changed their minds. Initially, only 8% of her readers believed that it paid to switch doors. Now 56% agree with Marilyn. And at universities the percentage of disbelievers drops from a majority of 65% to a minority of 29%. Some still stubbornly cling to their mistaken opinion, including the scientist who writes, "I still think you are wrong. There is such a thing as female logic".

Defying Common Sense

A Global Debate

Never before had people discussed a mathematical problem so vehemently and on such a large scale. Even among mathematicians, where one might least expect it, emotions were running high. One mathematician wrote to Marilyn, "There is enough mathematical illiteracy in this country, and we don't need the world's highest I.Q. propagating more. Shame!" How could a mathematical issue provoke such fierce public debate?

Marilyn vos Savant was not the first to write about the Monty Hall dilemma. As early as 1975, the American statistician Steve Selvin (1941–), inspired by Monty's game show, submitted two entertaining letters about the dilemma to the *American Statistician* journal. As several of his colleagues disputed that switching doors was the best strategy, he added a formal proof to the second letter, and there it ended.

When in September 1990 Marilyn wrote her first column on the Monty Hall dilemma, the situation was completely different. Only professional statisticians and a few mathematicians read the *American Statistician. Parade* magazine, on the other hand, had a readership of millions and millions. And while Steve Selvin may be well known among his colleagues, Marilyn, like Monty, was a national celebrity. In short, if Marilyn's column hadn't been so widely read and Monty's game show hadn't been so popular, the dilemma would never have provoked such a massive reaction.

But why did so many people erroneously believe that Marilyn was wrong? And why were they so strongly convinced that they themselves were right? In the *New York Times*, the mathematician Persi Diaconis commented on the dilemma: "I can't remember what my first reaction to it was, because I've known about it for so many years. [...] But I do know that my first reaction has been wrong time after time on similar problems. Our brains are just not wired to do probability problems very well. So I'm not surprised there were mistakes".

In other words, our first response is often mistaken because our brains are not equipped to solve probability problems. Maybe this explains why so many people were mistaken, but it does not explain why so many people stubbornly stuck to their mistake.

Deceived By Intuition

When Persi Diaconis speaks of his "first reaction" he is not talking about his logical or mathematical capacities; he is talking about his intuition. We often rely on our intuition, such as when we encounter a new problem and don't know how to handle it, or when we simply don't have enough time. Instead of analyzing the problem and deliberating between alternative solutions, intuition takes a shortcut. It plunges into the storehouse of our experience and looks for a familiar, similar-looking problem. This problem serves as a kind of analogy or model for the new problem, which is then "solved" in the same way.

For practical purposes, this shortcut works most of the time, *except* in situations where chance is involved. As we have seen in the previous chapters, our intuition is notoriously unreliable when it comes to solving probability problems.

Suppose someone asks you, "What is more likely, throwing at least one six in six rolls of a die, or throwing at least two sixes in twelve rolls of a die?" Your intuition will probably tell you that both outcomes have the same probability. Why is that? Because through experience we are familiar with many similar looking situations where the principle of proportionality does apply. For instance, a train that departs once every hour and a train that departs twice every two hours have the same frequency. But chance events don't behave like trains! The counter-intuitive but correct solution is that the former is more probable than the latter.

The Monty Hall Illusion

The Monty Hall dilemma has one of the most counter-intuitive solutions of all probability problems, if not the most counter-intuitive of all. Solving it, so it seems, doesn't require any mathematics. All you need is common sense. Two identical doors, one with a car behind it and the other with a goat behind it. How could switching doors make any sense? The intuitive answer that the chances are fifty-fifty appears self-evident. And yet it is wrong. How can a wrong answer seem so evidently right?

The dilemma closely resembles a situation that is basic to our very understanding of probability and randomness. Suppose there are no other possible outcomes than the ones given and the possible outcomes exclude each other. Then if the different options are symmetrical, the possible outcomes are equally probable.

Just as the outcomes of flipping a fair coin have equal probability because the two sides of the coin are symmetrical, so too do the two options of the Monty Hall dilemma have equal probability because of the apparent symmetry between the two doors.

So the dilemma is "solved" by invoking a familiar principle that not only often works, but even underlies our very understanding of probability. The principle of symmetry is part of the "wiring of our brains," so to speak. That is why the intuitive solution — the illusion that both doors have equal probabilities — seems self-evident. No explanation is needed. It's common sense.

Breaking the Spell

Nothing takes as much effort as letting go of an illusion disguised as common sense. This also explains the vehemence of the reactions. Instead of concentrating on the problem, many of the critics directed their criticism at Marilyn herself. Not only was she wrong, she had defied common sense. She had denied an elementary principle of probability! The fact that she was an intelligent female and a celebrity only added to the sarcasm and indignation. In her column of February 17, 1991, Marilyn herself wisely writes, "When reality clashes so violently with intuition, people are shaken".

How can we combat such a strongly rooted but erroneous intuition? One way is through experience. Simulate the game, repeat it many times and see what happens. Proof did not convince Marilyn's critics, nor did it convince Paul Erdős, but experiment did. Seeing is believing.

Another way is through a thought experiment in which the misleading model of our intuition is replaced by an appropriate one. Imagine, as Marilyn herself proposed in her first column, that there are not three identical doors but a million. Behind one of these doors is the car, while behind each of the other doors is a goat. You choose a door. Obviously, your chance of picking the winning door is only one in a million.

Next, the assistant opens the other doors one by one — each time a losing door — except for one door. What will you do, switch or stay? Do you believe that, of all the million doors, you picked the winning door right away? Or that your chance of winning the car is now fifty-fifty? Of course you don't. The car is behind the other door with an overwhelming probability of 999,999 in 1,000,000. You'll probably switch doors without thinking twice. And that's exactly how it also works with three doors.

In the initial situation of three identical doors, one arbitrary door is locked — the one chosen by the contestant — as is the door hiding the car. In this way an asymmetry is introduced into what appears on the surface as a symmetrical set-up of two identical doors, one door hiding the car, and the other a goat. So on the face of it, it seems to be no different than flipping a fair coin, while some in-depth analysis reveals otherwise.

As the Monty Hall Dilemma demonstrates, the capricious ways of chance often lead to surprising outcomes, but it is difficult not to get lost along the way. Intuition is a poor guide. Hopefully this book has served you as a more reliable guide in exploring the counter-intuitive logic of chance.

We will close with ten problems, challenging you to find your own way now. The problems are arranged from easy to difficult and follow the same order as the chapters in the book. Except for the last problem. This is the last of the fourteen propositions Christiaan Huygens discusses in his treatise *On the Calculus of Games of Chance*. It would be most fitting to end by quoting his very words: "May these problems serve the reader for practice and recreation!"

6 Ten Challenging Problems

Problems

Receiving the Nobel Prize Twice

In the introduction you read how Linus Pauling argued, upon being awarded the Nobel Prize for the second time, that the probability of winning the Nobel Prize a second time was much greater than that of winning the Nobel Prize for the first time and therefore much less special. Pauling was joking, but where exactly is the error in his argument?

A Medieval Commentary on Dante

In his masterpiece *The Divine Comedy* (*Divina Commedia*), the famous Italian poet Dante (1265–1321) describes how, when he climbs the mountain of Purgatory, the shadows of the deceased gather round him like the spectators at a game of Zara.

In his commentary on *The Divine Comedy*, the Italian scholar Jacopo della Lana (1290–1365) gives a description of the popular dice game Zara. Della Lana notes that sum 3, sum 4, sum 17, and sum 18 are ignored in Zara. These outcomes are too rare and therefore would make the game too long-winded. According to della Lana, all four combinations can only be thrown in one way. Was the Italian scholar right?

Too Good to Be True

The casino game of craps is still hugely popular in the US. In ancient times, all kinds of dice games were popular. Suppose a friend of yours proposes playing a dice game with two dice. If he or she rolls a two or a five you have to pay him or her a dollar; otherwise, your friend will have to give you a dollar. Clearly, the odds are in your favor. Or is there a catch?

An Aborted Tennis Match

Imagine Primo and Secondo playing tennis again. The two rivals are evenly matched. The first one to win six games gets 80 golden ducats. But bad luck strikes again. This time the match has to be aborted when Primo is leading 4–3. The prize money will be divided between the two players, but how does one assign each player his fair share?

Chuck a Luck!

Chuck-a-Luck is a game of chance traditionally played at carnivals in the American Midwest. It is played for entertainment or charity. A metal cage resembling a wire-framed birdcage, shaped more or less like an hourglass, contains three large dice. You can bet on one of the numbers 1 to 6. The usual bet is one dollar. The dealer spins the cage and eventually the three dice land at the bottom.

If your number appears on one of the dice, you win one dollar. If your number appears on two of the dice, you win two dollars, and if your number appears on all three dice, you win three dollars. In all three cases you'll also get your one dollar back. In all other cases you lose your one dollar.

Chuck-a-Luck *seems* to be an attractive game. A lot of people think that because their number — one in six — can match each one of the three dice, they have an even chance of winning. Besides, you can only lose one dollar, but win up to three! What do you think, is your chance of winning really 50%? And what is the expected value of this game, that is, the average payout you get per game in the long run?

It's Frank!

Suppose you meet some friends for lunch in a nearby restaurant. You talk about the things friends talk about when the conversation shifts to Frank, a friend you haven't talked to or seen in a long time. Nobody really knows how he's doing. You promise to call him soon. When you get home, you decide to call Frank right after dinner. But then your phone rings. It's Frank!

Could this really be a coincidence? To find the answer to this question you have to put this one event in its proper perspective. How many times over the years have you and your friends talked about Frank? And what are the chances he'll call you on a particular day? Make some rough estimates. What do these estimates tell you about coincidences like the one above?

Lucky Numbers

Say you're not a chronic gambler but when you go to the casino with friends you're glued to the roulette table. You've noticed that the ball often stops twice in a short time on the same number — as if, as they say, every day is someone's lucky day. If only you knew which number to pick in advance… What are the chances of the ball stopping twice on the same number in the span of eight rounds?

Tracking Down Terrorists

Little Brother is a novel by Cory Doctorow, written for and about young adults. The novel was published in 2008 and is freely available for download on the internet (http://craphound.com/littlebrother/download). It recounts how San Francisco turns into a police state after a terrorist attack on the San Francisco-Oakland Bay Bridge.

Seventeen-year-old Marcus Yallow and his crew, who are in the wrong place at the wrong time, are arrested, interrogated, and tortured for several days by the Department of Homeland Security. At a certain point, the main character gives a math lesson about the sense — and above all the nonsense — of giving up your liberty for a false sense of security.

Say the government has some data mining software to identify terrorists. It sifts through everyone's data about pin transactions, phone calls, public transport, and the like in the city. The software picks out terrorists 99% of the time, while innocent citizens are also identified 99% of the time. These high percentages look pretty impressive. And yet this anti-terrorism project is doomed to fail. Why is that?

Monty Hall 2.0

A TV producer came up with a new concept for the Big Deal. There are three identical closed doors on stage, but now a car, car keys and a goat are randomly placed behind the three doors. So behind each of the three doors is either the car, or the keys, or the goat.

Also, instead of one contestant, there are now two contestants. These contestants are on the same team. Each of them is given the chance to open and close two of the three doors while their teammate is backstage and cannot see or hear anything. They win the car when the first contestant opens the door with the car behind it and the second contestant opens the door with the keys behind it.

If they each open two doors at random, the probability of the first candidate finding the car is 2/3, and the probability of the second candidate finding the keys is also 2/3. So the chance that they win the car is $2/3 \times 2/3 = 4/9$ or 44%. However, the two candidates can agree on a strategy in advance. What is the greatest possible chance of winning the car? Can you think of a strategy that gives the two candidates a maximum chance of winning?

Christiaan Huygens' Game

Imagine that you were born in the seventeenth century and Christiaan Huygens is a good friend of yours. You are invited for dinner at the modest family estate of *Hofwijck*. To pass the evening Christiaan proposes playing a dice game.

The game he has in mind is played with two dice. The dice are rolled in turns. You win as soon as you throw six pips, while he wins as soon as he throws seven. Because the odds are in Christiaan's favor, you start. Who has a better chance of winning this game, Huygens or you?

If you approach this problem in a direct way, you'll get an infinite series of fractions: the probability of winning on the first roll, the probability of winning on the second roll, and so on. There is a shortcut that avoids calculating an infinite sum. Can you find it?

Solutions

Receiving the Nobel Prize Twice

Of course, the Nobel Prize Committee does not select candidates from a pool of former winners, but candidates who excel in a particular field — candidates who, for example, have made contributions to chemistry or to world peace. The pool of possible candidates is therefore much larger than the small pool of people who have already received the Nobel Prize once and are still alive.

A Medieval Commentary on Dante

Jacopo della Lana was right about the sums of 3 and 18. With three dice, both can only be rolled in one way. But he was wrong about the sums of 4 and 17. He writes that the sum of 4 can only be thrown as the combination of one two and two ones, but he overlooks that this combination can be achieved in three ways: 1 1 2, 1 2 1 and 2 1 1. The same goes for the sum of 17.

Too Good to Be True

At first glance, the odds seem to be 2 to 1 in your favor, but in fact the odds are in your friend's favor! The probability of throwing a two or a five with two dice is easy to find by first calculating the complementary probability of throwing neither a two nor a five. That probability is equal to $4/6 \times 4/6 = 16/36$, which is 4/9. So the probability of your friend throwing a two and/or five is 5/9. The odds are therefore 5 to 4 in favor of your friend.

An Aborted Tennis Match

A fair distribution of the prize money reflects each player's chances of winning the match if it were to be continued. Imagine that it is. Either Primo makes it 5–3 or Secondo ties the score to 4–4. At 5–3 Primo would be entitled to 70 ducats and Secondo to 10 ducats, as we calculated earlier. At 4–4, each player would of course be entitled to half the prize money, 40 ducats. So Primo is entitled to $1/2 \times 70 + 1/2 \times 40 = 55$ ducats. The remaining part of the prize money, 25 ducats, goes to Secondo.

Alternatively, we can also use Pascal's arithmetical triangle. Since a maximum of four games are to be played to finish the match, we need the fourth row of the triangle: 1 4 6 4 1. Secondo lacks three games, Primo two games. So the odds are $1 + 4 + 6 : 4 + 1$, that is 11 to 5 in favor of Primo. Primo therefore gets 55 out of 80 ducats and Secondo 25.

Chuck a Luck!

Let's calculate the complementary probability first. The probability that a single die won't show your number is 5/6. Using the product rule, it is now easy to calculate the complementary probability that none of the three dice will show your number: $5/6 \times 5/6 \times 5/6 = 125/216$. So the probability of at least one die showing your number is $1 - 125/216 = 91/216$. That's only 42.1%.

So the odds aren't even at all. However, you can still win up to three dollars while losing at most one. As Pascal has shown, we must weigh profits and losses on the basis of their probabilities. So what about the expected value of the game?

There are 75 possible outcomes where only one die matches your number (three times $1 \times 5 \times 5$); there are 15 possible outcomes where exactly two dice match your number (three times $1 \times 1 \times 5$); and there is only 1 possible outcome where all three dice match your number ($1 \times 1 \times 1$). There are no matches to your number in any of the remaining 125 possible outcomes of the dice.

If there is only one match you win 1 dollar, if exactly two dice match your number you win 2 dollars, and if all three dice match your number you win 3 dollars. In all other cases you lose your 1 dollar.

So the expected value is $(75 \times 1 + 15 \times 2 + 1 \times 3 - 125 \times 1)/216 = -0.08$ dollar. In other words, in the long run you'll lose an average of 8 dollar cents per game. That's about three times the house edge of a casino. So Chuck-a-Luck isn't as innocent a game as people think.

It's Frank!

You think of someone you haven't spoken to or seen in a long time, and, guess what, he or she calls you the same day. Or you think of someone, and the same day you happen to bump into him or her. With just a pencil and a scrap of paper you can make a rough estimate of the probability that something like this will happen.

Take Frank. Say, you've discussed him with your friends about 50 times in the last few years. Now it's clear he's not married to his phone. Let's say the probability of him calling you on any given day is about one in a hundred. So the complementary probability that he won't call you on a certain day is 99/100, and the probability that he won't call you on any of the fifty days you've discussed him with your friends is therefore $(99/100)^{50}$.

Now it's easy to calculate the probability of him calling you on at least one of the days you discussed him. This probability is $1 - (99/100)^{50} = 0.395$ — a surprising 39.5%! So the coincidence is not as amazing as you might think. These things just happen, simply by chance.

Lucky Numbers

Every fervent player knows that at times the roulette wheel seems to have a preference for certain numbers. They fall more often and in a relatively short time. Are these numbers lucky numbers? Is there a hidden power called Luck or is it all just coincidence?

In fact, it's the birthday problem in disguise. A certain number falling twice in eight rounds is mathematically equivalent to two of eight living creatures sharing their birthdays on some planet where a year lasts 37 days. Both are therefore calculated in the same way, by calculating the complementary probability first.

The number of possible outcomes of the roulette wheel with all eight numbers being different is $37 \times 36 \times \ldots \times 31 \times 30$. Dividing this number by the total number of possible outcomes, that is 37^8, gives you the complementary probability. Subtracting this number from 1 will give the probability that a number falls at least twice in eight rounds: $1 - 37 \times 36 \times \ldots \times 31 \times 30/37^8 = 0.56$. In other words, the probability is a surprising

56%. Someone may be very lucky at one of the many roulette tables in the casino. However, the chances that you are that person are very slim ...

Tracking Down Terrorists

Terrorists are relatively rare. Suppose there are 20 terrorists in a population of 20 million. That's one in a million or 0.0001%. Each of the 20 terrorists will most likely be picked out by the software program. Of the 20 million citizens, 99% will be marked as innocent and only 1% will be falsely classified as terrorists. But 1% of 20 million is still 200,000 citizens.

In other words, for every terrorist, an average of about 10,000 innocent citizens would have to be arrested and interrogated, even though the software is 99% accurate!

Monty Hall 2.0

Since the first contestant has a 1/3 chance of not opening the door with the car behind it, there is no strategy that gives a more than 2/3 chance of success.

Let's say the two contestants are the twins Pam and Sam. They've drawn lots and Sam is the one who gets to begin. Let's call the doors 1, 2 and 3, and since it doesn't matter which door Sam chooses first, let's say the twins agree that he will open door 1 first.

When the car is behind door 1, Sam has done all he needs to do. But what if it's the goat or the car keys? The twins agree that if it is the keys, Sam will open door 2 and if it is the goat, he will open door 3.

The car (C), the goat (G), and the keys (K) can be arranged behind the three doors in six possible ways: CKG, CGK, KCG, KGC, GCK, and GKG. In two of the six arrangements (KGC and GCK) Sam will not open the door with the car behind it. So the chance that the twins will already lose on their first attempt is 2 out of 6, or 1/3.

As we observed, the greatest possible chance of winning the car was 2/3. So Pam will have to open the winning door, the door with the keys behind it, in each of the remaining four arrangements. Otherwise, the twins' strategy will give them a less than 2/3 chance of winning. Can we find a strategy that guarantees Pam will open the winning door?

We can. The four remaining possible arrangements are CKG, CGK, KCG and GKG. Pam opens door 2 first. When the keys are behind door 2, the twins will have won the car. But what if it's the car or the goat?

That's simple. If it is the car (KCG) Pam has to open door 1, and if it is the goat (CGK) Pam has to open door 3.

In short, if Sam opens the door with the car behind it, this strategy guarantees that the twins will win the car. Otherwise, the game will already be lost in the first round. And that chance is 1/3. So the above strategy gives them a 2/3 chance of winning the car!

Christiaan Huygens' Game

When the German child prodigy Carl Friedrich Gauss (1777–1855) — one of the greatest mathematicians ever — was still in primary school, the teacher one day instructed the pupils in the classroom to add up the numbers 1 to 100. Convinced that this would keep them busy for a long time, he was about to take a nap when, after a few seconds, young Gauss came up with the solution. Instead of adding up the numbers one by one, as his classmates did, he reasoned as follows. Adding 1 to 100 equals 101, adding 2 to 99 also equals 101, and so on. In short, the sum of the numbers 1 through 100 is equal to fifty times 101, which is 5050.

For the game proposed by Christiaan Huygens we need something similar so that we don't have to add up an infinite series of fractions. It's clear that when you both throw the dice once without winning, you are in exactly the same position as at the start of the game. As long as neither one of you succeeds in winning, the game actually starts over and over again.

Now let's call the unknown probability that you win p. You win the game if either (1) you win on the first roll or (2) neither one of you wins on the first roll and the game starts over again and again until you finally win. So if we know the probabilities of (1) and (2), we also know p. It's the sum of the probabilities of (1) and (2).

First, the easy part. As you can check for yourself, the probability that you will throw six pips is 5/36 and the probability that Christiaan will throw seven pips is 6/36. So the probability of (1) is 5/36.

The probability of (2) is a little harder to find. When we take the complementary probabilities of throwing respectively six and seven pips and use the product rule, we get the probability that neither one of you wins on the first roll: $(1 - 5/36) \times (1 - 6/36)$. Since the game then actually starts all over again, your probability of winning will also be p again. So the probability of (2) is $(1 - 5/36) \times (1 - 6/36) \times p$.

Now, you win if either (1) or (2) is the case. The probability p is therefore equal to the sum of the probability of (1) and the probability of (2): $p = 5/36 + (1 - 5/36) \times (1 - 6/36) \times p$. If you solve p from this equation, you'll find that $p = 30/61$. So the odds are actually 30 to 31 in favor of Christiaan Huygens, even though you roll the dice first.

Epilogue. The Two-Edged Sword

When I started writing this book, my plan was to offer the reader a kind of sightseeing tour through the surprisingly counter-intuitive world of chance. However, it soon became clear that this mission could not be accomplished in an intellectually satisfactory way without answering the question of *why* our intuition deceives us time and again.

I became more and more convinced that psychological explanations didn't get to the heart of the matter. Psychology can play a role, but at the base of all probabilistic fallacies invariably lies the seemingly correct, but improper application of elementary probability principles. Let's illustrate this with the most famous and persistent probabilistic misconception, the gambler's fallacy.

In the psychology of judgement the gambler's fallacy is often explained as a cognitive bias that is rooted in a psychological mechanism called the *representativeness heuristic*. This mental shortcut tells us that if *A* belongs to class *B* then *A* is representative of *B* — that is, *A* shares essential features with *B*.

Applied to random processes the representativeness heuristic causes us to think that short sequences of random events are representative of longer ones. As a consequence we intuitively expect that deviations from average should balance out in the short run. Tversky and Kahneman coined this erroneous belief *the belief in the law of small numbers*. Accordingly, our intuition tells us that after a sequence of heads the probability of tossing tails increases. In short, we commit the gambler's fallacy.

The problem with applying the representativeness heuristic to long-term processes is that in reality we do exactly the opposite: we don't project the long-term behavior of processes onto their short-term behavior, on the contrary, we intuitively extrapolate our short-term experiences to their long-term behavior. We project the known onto the unknown. That's why we keep underestimating the increase of exponential processes such as the spread of covid-19.

The gambler's fallacy is not the result of a psychological mechanism wired in our brains but is rooted in our intuitive understanding of the law of large numbers itself. As Bernoulli observed, even an "idiot" with no mathematical training whatsoever knows this law. But what we don't know is how large the numbers to which the law applies actually are. As a result we intuitively apply the law to familiar numbers, the numbers in the "number domain" of our daily lives.

In reality the law applies to numbers which are far beyond the reach of our intuition. As we have seen in the case of tossing a coin, "in the long run" means "in the *very long* run". In other words, our intuition literally takes a shortcut.

In the same way, most other probabilistic fallacies can be explained as the misapplication of elementary principles of probability. So if we want to unravel the counter-intuitive logic of chance, we first need some basic understanding of the rules of probability. I figured that an excursion into the early history of probability would be an suitable way to provide the reader with a minimum of mathematical baggage. Again, I had to navigate deeper waters than I had expected.

On the surface, probability is one of those familiar concepts that appear to have always been part of our daily lives. But beneath the surface, it turns out to be one of the most problematic concepts in the history of mathematics.

Quite surprisingly, the ancient Greeks and Romans did not yet have a notion of probability. What's more, it is not until the seventeenth century that we encounter the concept of probability in its modern sense. How, then, did mathematical probability emerge?

The answer to this question not only is of historical importance, but can also shed light on another long-standing and hotly debated issue: what defines our concept of probability and how does mathematical probability relate to other concepts of probability?

So far, none of these issues has been resolved. Solutions have been proposed, but not one of them has proven to be conclusive or generally accepted. It is rather strange, to say the least, that a branch of mathematics derives its name from a concept of which it knows neither the origin nor the meaning!

My classical training allowed me to take a fresh look at these long-standing issues. An exhaustive examination would, of course, be contrary to the spirit of this book. In this epilogue I will therefore summarize some of my findings and ideas, and hopefully open up new perspectives on these old questions.

It is well known that the roots of probability theory lie in gambling with dice. Until recent times, gambling with dice was a popular pastime all over the world. Ancient dice have been found from Mesopotamia and India to Egypt, Greece, and Rome. In the first century BCE, Cicero wrote that nothing is as unpredictable as the outcomes of dice. In other words, the die is the archetype of a random number generator.

Most dice games were played with more than one die, usually three. As every gambler is sure to have observed, some sums of the three dice tend to occur more frequently than others. So there must be some system to the random behavior of dice. This observation led to the disclosure of the mathematical principles of dice games. Let's start with the key principle.

Let's call all ordered combinations of pips that add up to a single sum *favorable outcomes*, and all other combinations *unfavorable outcomes*. Then each sum tends to occur less or more frequently according to the ratio of its favorable and unfavorable outcomes. Let's call this principle *the proportionality principle*. It was presumably already known in ancient times, but our first written testimony dates only from the thirteenth century CE.

The principle is quite straightforward, but it only applies as long as we play with fair dice. A fair die is a die that has no preference for one of its six possible outcomes. This property has its roots in another mathematical principle.

An ideal die has the shape of a cube. Since all the faces of a cube are identical, a cube still looks the same no matter which face it lands on. In mathematical terms, a cube is symmetrical under rotation. As a result all outcomes of an ideal die will tend to occur with the same frequency. Let's call this principle *the symmetry principle*. It must already have been known in ancient times, as the large number of loaded dice that have been excavated reveals.

These two principles enable us to precisely calculate for each sum of the dice how more or less frequently it tends to occur than any or all of the other sums. Earlier we called this ratio *the odds*.

Since tendencies only show up in the long run, how often each sum relatively *tends to occur* is therefore reflected in its relative frequency *in the long run*. In other words, the odds will emerge in the long run. This is exactly the intuitive notion of the law of large numbers that Bernoulli wrote about. Let's call it *the law of the long run*.

The law is often referred to as the *empirical* law of large numbers, implying that we know the law from experience. The law, however, applies to numbers that are much larger than the numbers we deal with in daily life. In fact, the law is inherent to our concept of randomness, as can be argued as follows.

If, in the long run, the relative frequencies did not "stabilize" around the calculated odds, this would create an internal contradiction. The proportionality principle would no longer be valid and our concept of randomness would become meaningless. This was seemingly the case with the gamblers who asked Galileo for advice, because on the basis of their long experience they came to different conclusions than their calculations indicated.

In short, we possess an intuitive concept of random events that is rooted in the possible outcomes of dice. Its rules are the symmetry principle, the proportionality principle, and the law of the long run. These principles link the proportion of favorable outcomes of each event to its relative frequency in the long run. Let's call this unarticulated property the *accidentality* of a random event. For instance, when you roll two dice, the sum of 3 has an accidentality of 1/12.

Jacob Bernoulli made two groundbreaking contributions. He proved the mathematical counterpart of our intuitive law of the long run, the law of large numbers. He also formally introduced what we called the accidentality of an event as a value between 0 and 1. Only he called it *probability*. Why?

The term probability was already in use, but in those days it meant something quite different. It was used with respect to opinions and judgements and expressed a degree of approval or belief. So how can this qualitative concept suddenly become a mathematical quantative concept?

Bernoulli most likely adopted the term from the *Port-Royal Logic*. The *Port-Royal Logic* was written by Antoine Arnauld in response to the trendy Skepticism of his time. Arnauld shared Descartes' quest for certainty in science, but admitted that in the practice of life absolute certainty, and even practical certainty, is unattainable most of the time. Instead, one should choose the *most probable* judgement.

The probability of a judgement depended on what Arnauld called the circumstances, that is, its evidence. However, by definition, there is no evidence for a judgement about a chance event. After all, chance events are unpredictable. The final chapter of the *Port-Royal Logic* shows us how we can still deal with chance events.

One could say that a chance event has its own degree of uncertainty, so to speak. For instance, only one in a million people is killed by lightning. In the *Port-Royal Logic* this uncertainty of an event is *by analogy* called the *probability of the event.* Analogies, as is well known, can help make new concepts meaningful to students.

Bernoulli had studied the *Port-Royal Logic* thoroughly, as is evident from Part Four of his *Ars Conjectandi*. He adopted the new use of the term *probability* — popularized by the success of the *Port-Royal Logic* — and formally introduced it into mathematics. The term *certainty* in probability theory also goes back to Bernoulli and ultimately derives from the *Port-Royal Logic,* that is, from the debate between Skeptics and Cartesians like Arnauld.

But it wasn't Arnauld who was behind this new use of the term probability. As we have seen, the spirit of mathematics that pervades the last chapter of the *Port-Royal Logic* and its call for the reader to lead a devout life unmistakably bear Pascal's stamp. There is little doubt that it was Pascal who took the term probability out of its epistemic context in the previous chapters with the intention of clarifying his mathematical concept of expected value, the guiding principle when it comes to chance events.

So, as with many technical terms, the new concept did not emerge; it was born, and likely none other than Blaise Pascal was its intellectual father. As a consequence, all inquiries into the mathematical concept of probability that depart from the assumption that probability somehow evolved from a qualitative concept into a quantitative concept, either gradually or abruptly, are doomed to fail. Mathematical probability and probability in its original sense are *not* branches of the same tree at all. Mathematical probability only acquired its name by means of an analogy.

Naming a new scientific concept after an existing one is common practice, but sometimes this practice can prove to be a two-edged sword. The original concept of probability is often called *epistemic probability* (after the ancient Greek word *episteme*, "knowledge"), while the mathematical concept is named *aleatory probability* (after the Latin word *alea,* "die"). This suggests that the two concepts are somehow siblings.

Bernoulli himself was the first to fall victim to the misconception that the two concepts are somehow related. He extended the use of the term probability to statistical data. That is, data that can also be expressed as relative frequencies but cannot be known beforehand. For instance, what is the probability that a young man will die earlier than an old man? He believed that this extension was justified by the law of large numbers.

His extension of the mathematical concept to statistical data was the basis for constructing a calculus for judgements under uncertainty. The title of Part Four of the *Ars Conjectandi* speaks about "applications of probability to civil, economical and moral matters".

In the *Port-Royal Logic* certainty had been the criterion for science, while probability had been the criterion for judgements in practical life. Quite likely, Bernoulli, who had thoroughly studied the works of René Descartes, wanted to provide an objective method for practical judgements, as Descartes had done for science.

His quantification of epistemic probability was inspired by the new aleatory concept of probability. But, as Henri Poincaré already warned, good judgement should not be replaced by numbers. The two concepts differ categorically, the one dealing with events, which are subject to mathematical calculations, and the other dealing with judgements, which are subject to rational debate.

Today, there are even more concepts of probability, creating a "Babylonian confusion", as the great statistician Leonard Savage once wrote. Can we find a way out? As we have seen in the case of aleatory and epistemic probability, they should not be treated as close relatives or relatives at all.

Moreover, the formalization of epistemic probability presupposes that there is some criterion inherent to the concept itself. However, the dynamics of the concept of probability vary greatly according to its domain, i.e. the field to which it is applied. For instance, the dynamics of probability in court are very different from the dynamics of probability on the trading floor. And even in court cases differ greatly.

Bernoulli's aim was to "measure the probability of things as precisely as possible, to always be able to choose the best and safest road in our judgements and actions". "The wisdom of the philosopher and the prudence of the statesman," he adds, "depend on this".

We are living in unprecedented times. A calculus of epistemic probability that shows us the best and safest road would be most welcome, but I'm afraid that the world would be better off if, instead of pursuing an unattainable ideal, the philosophers and statesmen of our time were to invest their energy in exercising good judgement.

References and Further Reading

Part One

General Literature

A once authoritative but now outdated work on the history of probability theory is Isaac Todhunter (1865), *A History of the Mathematical Theory of Probability from the time of Pascal to that of Laplace*, Cambridge/London: Macmillan. A popular and readable classic, but unfortunately also outdated and not always correct is Florence N. David (1962), *Games, Gods and Gambling*, New York: Hafner Publishing Company. More recent is Anders Hald (2003), *A History of Probability and Statistics and Their Applications before 1750*, Hoboken, New Jersey: John Wiley & Sons, an excellent book, but mainly aimed at a mathematically advanced readership. Another excellent book, but also aimed at a mathematically advanced readership is Prakash Gorroochurn (2012), *Classic Problems of Probability*, Hoboken, New Jersey: John Wiley & Sons.

The Silence of Antiquity

A comprehensive book on stochastic thinking in antiquity is Robert Ineichen (1996), *Würfel und Wahrscheinlichkeit*, Heidelberg: Spektrum Akademischer Verlag. An acclaimed book on the history of gambling is David G. Schwartz (2006), *Roll the Bones. The History of Gambling*, New York: Gotham Books (not the "Casino Edition"!).

A discussion and overview of the various theories on the apparent absence of a probability calculus in Greco-Roman times can be found in Ian Hacking (2006), *The Emergence of Probability*, Cambridge, Cambridge University Press, pp. 1–10. See also Daniel Garber and Sandy Zabell (1979), "On the Emergence of Probability," *Archive for History of Exact Sciences* 21, p. 49; and Ineichen (1996), p. 138.

Aristotle's ideas about chance can be found in *Physics*, Book Two, chapters 4–6 and *Metaphysics*, Book Eleven (K), chapter 8. For Epicurus's opinion on chance, see his *Letter to Menoeceus*.

On the concept of chance in Cicero, see David (1962), pp. 24–5. The Latin text of Cicero's *De divinatione* with an English translation can be found in Cicero (1923), *On Old Age. On Friendship. On Divination.* Trans. by W. A. Falconer, Cambridge, Massachusetts: Harvard University Press (*Loeb Classical Library* No. 154).

Most ancient texts and their English translations are also available online. Some well-known collections are the *Perseus Digital Library*, *The Latin Library*, *Lacus Curtius* and *Wikisource*.

The German Wikipedia-article "Karneades von Kyrene" provides extensive biographical information about Carneades. The most important ancient source for biographies of ancient philosophers is Diogenes Laërtius, *Lives and Opinions of Eminent Philosophers*. The anecdote about Carneades's housekeeper feeding Carneades while he was reading can be found in Valerius Maximus, *Memorable Doings and Sayings* Book 8, chapter 7. The *Gambler's Lament* is hymn 34 in the tenth book of the *Rigveda*.

A Medieval Ovid

Gambling in the Middle Ages is extensively treated in Schwartz (2006) *Roll the Bones*.

Ordericus Vitalis mentions gambling clergymen in his Church History (*Historia Ecclesiastica*, p. 550), cited by Joseph Strutt (1903), *Sports and Pastimes of the People of England*, Londen: Methuen, 1903, p. 246.

On the pious dice game of Bishop Wibold, see Richard Pulskamp and Daniel Otero (2014), "Wibold's Ludus Regularis, a 10th Century Board Game", *Convergence* (June), published on the website of the Mathematical Association of America (www.maa.org/publications/periodicals/convergence/wibolds-ludus-regularis-a-10th-century-board-game).

On the "probability" calculations in the pseudo-Ovidian poem *The Old Woman*, see David R. Bellhouse (2000), "De Vetula: A Medieval

Manuscript Containing Probability Calculations", *International Statistical Review/Revue Internationale de Statistique*, Vol. 68, No. 2 (August), pp. 123–136. The Latin text and English translation of the relevant passage can be found in this article.

The poem itself is discussed by Dorothy M. Robathan (1957), "Introduction to the Pseudo-Ovidian De Vetula", *Transactions and Proceedings of the American Philological Association*, Vol. 88, pp. 197–207.

On the infamous blunder of Leibniz, see Prakash Gorroochurn (2011), "Errors of Probability in Historical Context", *The American Statistician*, 65 (4), pp. 246–254; see also Todhunter (1865), p. 48. Leibniz's letter can be found in Gottfried W. Leibniz, *Die philosophischen Schriften*. Bd. 3, C.I. Gerhardt (ed.), Berlijn: Weidmann, 1887; (Reprint: Hildesheim/New York: Olms, 1978).

Petrarch's sarcastic comment on Ovid's authorship of the poem is mentioned in Pierre de Nolhac (1907), *Pétrarque et l'Humanisme,* Paris: H. Champion, 1.179. It can be found in the fourth letter of the second book of his *Letters of Old Age* (*Epistolae Seniles* II, 4, 12).

The Divine Mathematician

Galileo's famous words about the study of nature are from *The Assayer* (*Il Saggiatore*, 1623), which is generally considered to be a pioneering work of the scientific method.

All texts and documents relating to Galileo's trial by the Inquisition can be found in Maurice A. Finocchiaro (1989), *The Galileo Affair: A Documentary History*, Berkeley: University of California Press. Also in Maurice A. Finocchiaro (2008), *The Essential Galileo*, Hackett Publishing Company.

By far the best biography of Galileo is John L. Heilbron (2010), *Galileo*, New York: Oxford University Press.

The Italian text of *Sopra le Scoperte dei Dadi* can be found in Galileo Galilei, *Opere*, Florence: Barbera, vol. 8 (1898), pp. 591–594. The text is also available online at www.liceoagnoletti.it/public_html/files/Galileo_Galilei_sui_tre_dadiBarra.pdf.

An English translation is provided in David (1962), pp. 192–195. The quotation from the editorial preface to Galileo's collected works can also be found in David (1962), p. 67.

On the mistake made by Jacopo della Lana, see Robert Ineichen (1988), "Dante-Kommentare und die Vorgeschichte der Stochastik," *Historia Mathematica* 15, pp. 264–269.

The Gambling Scholar

An English translation of Cardano's autobiography is: Jean Stoner (2002), *Cardano, Girolamo, The Book of My Life*, New York: New York Review of Books.

An edition of the Latin text of Cardano's *Book on Games of Chance* is Massimo Tamborini (2006), *Girolamo Cardano. Liber de ludo aleae*, Milan: Franco Angeli. An English translation with biographical information and extensive commentary can be found in Ore Øystein (1953), *Cardano. The Gambling Scholar*, New York: Dover publications. A recent reprint of this English translation by Sydney H. Gould is *Gerolamo Cardano, The Book on Games of Chance*, New York: Dover Publications, 2015. It should be noted that the translation is not always accurate and contains anachronisms.

Cardano's *Collected Works* (Lyon 1663) are available online at http://www.cardano.unimi.it/testi/opera.html.

An ample selection from the correspondence between Ferrari and Tartaglia can be found in chapter three of Øystein (1953). On the feud between Cardano and Tartaglia, see also Hal Hellman (2006), *Great Feuds in Mathematics. Ten of the Liveliest Disputes ever*, New York: John Wiley & Sons, pp. 7–25.

The Aristotelian background of Cardano's quest for equality was first uncovered by David Bellhouse (2005), "Decoding Cardano's Liber de Ludo Aleae", *Historia Mathematica* 32, pp. 180–202.

The Child Prodigy and the Counselor

There exist many biographies of Pascal. A classic one is Jean Mesnard (1962), *Pascal*. Paris: Hatier. By far the best biographical introduction to

Pierre Fermat is the German Wikipedia-article “Pierre Fermat”. See also Klaus Barner (2009), “Pierre Fermat. Sa vie privée et professionnelle,” *Annales de la Faculté des sciences de Toulouse: Mathématiques*, Tome 18, no. 2, pp. 119–135, online available on https://afst.centre-mersenne.org. On Pascal and the Chevalier de Méré, see Øystein Ore (1960), “Pascal and the Invention of Probability Theory,” *American Mathematical Monthly*, 67, pp. 409–419. On the popularity of gambling in the age of King Louis XIV, see Schwartz (2006), *Roll the Bones*, pp. 71–72.

On the history of the problem of the unfinished game (the problem of points), see Ernest Coumet (1965), “Le problème des partis avant Pascal,” *Archives Internationales d'Histoire des Sciences* 18, pp. 245–272, and Norbert Meusnier (2007), “Le problème des partis bouge… de plus en plus,” *Journal Electronique d'Histoire des Probabilités et de la Statistique* 3, no.1, pp. 1–27. Both articles are available from www.jehps.net.

The correspondence between Fermat and Pascal, Pascal’s letter to the Parisian Academy, as well as the *Traité du triangle arithmétique* can be found in one of the many editions of Pascal's Collected Works, e.g. Pascal (1954), *Oeuvres completes. Texte établi, présenté et annoté par J. Chevalier*, Parijs: Gallimard (Pléiade).

An English translation of the correspondence and the Treatise can be found in Richard Scofield (1988), *Great Books of the Western World. Pascal*, Chicago: Encyclopedia Britannica, pp. 447–487. The English translation provided in David (1962), *Appendix 4*, pp. 229–253 is not entirely flawless. An English translation of the *Pensées* can also be found in *Richard Scofield* (1988).

On Pascal and the problem of the unfinished game, see Anthony W.F. Edwards (1982), “Pascal and the Problem of Points,” *International Statistical Revue* 50, pp. 259–266, reprinted in Anthony W.F. Edwards (2002), *Pascal’s Arithmetic Triangle: The Story of a Mathematical Idea.* Baltimore: John Hopkins University Press, Appendix 1, pp. 138–150. For a mathematical treatment, see Hald (2003), “Chapter 5. The Foundation of Probability Theory by Pascal and Fermat in 1654.” and Gorroochurn (2012), pp. 20–38.

The Dutch Archimedes

The complete correspondence and the complete works of Christiaan Huygens are collected in Christiaan Huygens (1888–1950), *Oeuvres Complètes*, 22 volumes, Société hollandaise des Sciences: Den Haag: Martinus Nijhoff, also available on www.dbnl.org/auteurs/auteur.php?id=huyg003.

Volume I and II contain Huygens' correspondence up to 1657. Volume XIV is dedicated to Huygens' contributions to the field of probability theory. For the Dutch version of Christiaan Huygens' treatise on games of chance, see Wim Kleijne (1998), *Christiaan Huygens. Van Rekeningh in Spelen van Geluck*, Utrecht: Epsilon Uitgaven.

A lot of biographical information can be found in the prefaces and notes in the *Oeuvres Complètes*. Volume XXII contains a voluminous French biography by J.A. Vollgraff. An excellent recent biography is Hugh Aldersey-Williams (2020), *Dutch Light*, London: Picador. Another biography is C.D. Andriesse (2011), *Huygens: The Man Behind the Principle*, Cambridge: Cambridge University Press.

On Pascal and the problem of gambler's ruin, see Anthony W.F. Edwards (1983), "Pascal's Problem: The 'Gambler's Ruin'", *International Statistical Revue*, vol. 51, no. 1, pp. 73–79, reprinted in Edwards (2002), "Appendix 2", pp. 157–166. See also Hald (2003), "Chapter 6, Huygens and De Ratiociniis in Ludo Aleae, 1657"; and Gorroochurn (2012), pp. 39–48.

The Art of Thinking

A recent edition of the French text is Antoine Arnauld et Pierre Nicole (1992), *La logique ou l'art de penser*, Paris: Editions Gallimard. An English translation is Jill V. Buroker (1996), *Arnauld, Antoine, and Pierre Nicole. Logic or the Art of Thinking*, Cambridge: Cambridge University Press.

A classic work on the revival of ancient Skepticism in the 17th century is Richard Popkin (2003), *The History of Scepticism,* Oxford: Oxford University Press. On the influence of Descartes' philosophy on Arnauld, see Buroker (1996), pp. xx–xxii. Descartes' remarks are from his *Discourse on the Method* and his *Objections and Replies VII*; they are quoted in Popkin (2003), pp. 143–145.

On the use of the term *probability* before Pascal, see James Franklin (2001), *The Science of Conjecture. Evidence and Probability before Pascal*. Baltimore/London: The John Hopkins University Press.

On Cicero and the Latin term *probabile*, see John Glucker (1995), "*Probabile, Veri Simile,* and Related Terms," in: J.G.F. Powell (ed.), *Cicero the Philosopher: Twelve Papers*, Oxford: Oxford University Press; see also Franklin (2001), pp. 113; 126.

Pascal's intellectual contribution to the last chapter of the *Port-Royal Logic* is discussed in Anthony W.F. Edwards (2003), 'Pascal's Work on probability' in: *The Cambridge Companion to Pascal*, edited by Nicholas Hammond. Cambridge: Cambridge University Press, pp. 40–52.

The Golden Theorem

All of Jacob Bernoulli's work relating to probability can be found in volume three of Jakob Bernoulli (1975), *Die Werke von Jakob Bernoulli*, edited by Bartel L. Van der Waerden, Basel: Birkhäuser Verlag.

An English translation of the *Ars Conjectandi* is Edith D. Sylla (2006), *Jacob Bernoulli: The Art of Conjecturing. Together with Letter to a Friend on Sets in Court Tennis*, Baltimore: Johns Hopkins University Press.

The letter to Leibniz can be found in G.W. Leibniz, *Mathematische Schriften*, Vol. III, edited by C.I. Gerhardt, Halle, 1855 (reprinted 1960–1961, Hildesheim, Georg Olms; digitalized by Google). A German translation can be found in Bernoulli (1975), *Die Werke*.

An English translation of excerpts from Bernoulli's correspondence with Leibniz and excerpts from Part Four of the *Ars Coniectandi* is Bing Sung (1966), *Translations from James Bernoulli*, Department of Statistics, Harvard University, Cambridge, available on
www.matematica.ciens.ucv.ve/modelos/Descargas/ars_sung.pdf.

Biographical information can be found in Sylla (2006); David (1962), pp. 130–139; and in Karl Pearson (1978), *The History of Statistics in the 17th and 18th Centuries*, New York: Macmillan, pp. 212–230.

On Jacob's non-mathematical activities, see Fritz Nagel (2006), "Jacques Bernoulli l'Inconnu. Quelques aspects des activités non mathématiques d'un mathématicien," *The Electronic Journal for History of Probability and Statistics*, vol. 2/1b (November). On the feud between Jacob and his brother Johann, see Jeanne Peiffer (2006), "Jacob Bernoulli, Teacher and Rival of his Brother Johann," *The Electronic Journal for History of Probability and Statistics*, vol. 2/1b (November). Both articles are available on www.jehps.net.

On the law of large numbers, see Hans Freudenthal (1972), "The 'Empirical Law of Large Numbers' or 'The Stability of Frequencies'," *Educational Studies in Mathematics*, Vol. 4, No. 4, pp. 484–490; Henk C. Tijms (2012), *Understanding Probability*, Cambridge: Cambridge University Press, Cambridge, Part One, chapter 2; and Gorroochurn (2012), *Classic Problems*, p. 64. On the history of generating random numbers, see Deborah J. Bennet (1998), *Randomness*, Cambridge, Massachusetts: Harvard University Press, pp. 132–151.

Part Two

The Gambler's Fallacy

The historical background of the casino is treated in David G. Schwartz (2013), *Roll the Bones. Casino Edition*, Las Vegas: Winchester Books.

The gambler's fallacy is discussed in the Wikipedia article "Gambler's falllacy". One of the earliest descriptions of the fallacy can be found in Pierre-Simon Laplace (1796), *A Philosophical Essay on Probabilities*. The game of roulette at the Monte Carlo Casino on August 18, 1913 is probably the most famous example of the gambler's fallacy.

On the "collective psychosis" of the Italian people, see "No 53 puts Italy out of its Lottery Agony," *The Guardian*, Febuari 11, 2005; and "Lotto, estratto il 53 a Venezia. Vincite fra 600 e 800 milioni," *La Repubblica*, February 9, 2005; also see "Long-awaited lottery number comes up," *The Lottery Post*, February 10, 2005 (www.lotterypost.com/news/107381).

A short story dealing with the gambler's fallacy is Colin Bruce (2002), "The Case of the Gambling Nobleman", in: *Conned Again, Watson! Cautionary Tales of Logic, Math, and Probability*. New York: Perseus Books.

The belief in the law of small numbers is discussed in Daniel Kahneman (2012), *Thinking, Fast and Slow*. London: Penguin Books, pp. 109–118; pp. 419–432 (*Appendix A:* Amos Tversky and Daniel Kahneman (1974), *Judgement under Uncertainty: Heuristics and Biases*). The original article by Amos Tversky and Daniel Kahneman is "Belief in the Law of Small Numbers", *Psychological Bulletin* 76 (2), 1971. pp. 105–110.

Streaks and Runs

On long runs of the same outcomes, see Mark F. Schilling (1990), "The Longest Run of Heads," *The College Mathematics Journal*, 21(3), pp. 196–207; Mark F. Schilling (2012), "The Surprising Predictability of Long Runs," *Mathematics Magazine*, 85(2), pp. 141–149.

For the professor's assignment to his students, see (among many) Pál Révész (1978), "Strong Theorems on Coin Tossing," *Proceedings of the International Congress of Mathematicians*, Helsinki (1978), pp. 749–754; see also the readable article by Frank A. Martin (2009), "What were the Odds of Having Such a Terrible Streak at the Casino?", available on https://wizardofodds.com/image/ask-the-wizard/streaks.pdf.

An instructive computer simulation by S. M. Blinder (2011) can be found on http://demonstrations.wolfram.com/ConsecutiveHeadsOrTails.

The article that provoked so much consternation, is Thomas Gilovich, Robert Vallone and Amos Tversky (1985), "The Hot Hand in Basketball: On the Misperception of Random Sequences," *Cognitive Psychology* 17, pp. 295–314. Also, Thomas Gilovich (1991), *How We Know What Isn't So*. New York: The Free Press, pp. 11–20; Amos Tversky and Thomas Gilovich (2005), "The Cold Facts About the 'Hot Hand' in Basketball," in: Jim Albert, Jay Bennet and James J. Cochran (eds.), *Anthology of Statistics in Sports*, Philadelphia: Society for Industrial and Applied Mathematics, pp. 169–174.

Recently, mathematical support for the existence of the hot hand has been put forward, see George Johnson (2015), "Gamblers, Scientists and the Mysterious Hot Hand," *The New York Times*, October 18.

For an overview, see the article "Hot hand" on Wikipedia.

Chance Fluctuations

On Bill Gates promoting smaller schools, see Howard Wainer en Harris L. Zwerling (2006), "Evidence That Smaller Schools Do Not Improve Student Achievement," *The Phi Delta Kappa International*, Vol. 88, No. 4, December, pp. 300–303.

On random fluctuations in the rating of small hospitals and in the crime rates of small towns, see Howard Wainer (2007), "The Most Dangerous Equation," *American Scientist*, Vol. 95, No. 3, May–June, pp. 249–256, reprinted in: Howard Wainer (2009), *Picturing the Uncertain World*, Princeton, New Jersey: Princeton University Press, pp. 1–20.

On the insensitivity to sample-size in psychology, see Anna Mikulak (2013), "Psychological Scientists Call for Paradigm Shift in Data Practices", available on http://www.psychologicalscience.org/index.php/publications/observer/2013/july-august-13/psychological-scientists-call-for-paradigm-shift-in-data-practices.html.

Defying the Odds

The story of Maureen Wilcox is a famous one. See for example David J. Hand (2014), *The Improbability Principle: Why Coincidences, Miracles, and Rare Events Happen Every Day*, New York: Scientific American, p. 92.

On Evelyn Adams winning the New Jersey state lottery twice, see Robert D. McFadden (1986), "Odds-Defying Jersey Woman Hits Lottery", *The New York Times*, February 14; Stephen M. Samuels and George P. McCabe Jr. (1986), "More Lottery Repeaters Are on the Way", *The New York Times* (*Opinion*), February 27; and Gina Kolata (1990), "1-in-a-Trillion Coincidence, You Say? Not Really, Experts Find", *The New York Times*, February 27, 1990.

On the MacKriell family, see Hannah Fletcher (2008), "An easy birthday to remember: Couple have their third child on the same date", *The Times*, February 26, and "Couple gives birth to three children on the same day… 14 years apart", *The Daily Mail Online*, February 25, 2008. Also discussed by David Hand (2008), "Three children with the same birthday?", submitted on May 2 to the site *Understanding Uncertainty* (https://understandinguncertainty.org/node/92).

On Linsay Hasz biting on a rare pearl, see Avianne Tan (2016), "Washington Woman Finds Rare Purple Pearl Worth $600 in Her Food While Dining Out", *ABC News*, February 24, (https://abcnews.go.com/US/washington-woman-finds-rare-purple-pearl-worth-600/story?id=37163589).

The Law of Truly Large Numbers

The classic article on the law of truly large numbers is Persi Diaconis and Frederick Mosteller (1989), "Methods for Studying Coincidences," *Journal of the American Statistical Association* 84 (408), pp. 853–861.

The story of the woman who was hit by a meteorite is told in Justin Nobel (2013), "The True Story of History's Only Known Meteorite Victim," *National Geographic News*, February 20 (http://news.nationalgeographic.com/news/2013/02/130220-russia-meteorite-ann-hodges-science-space-hit/). See also John C. Hall (2007), "Hodges Meteorite Strike (Sylacauga Aerolite)," *Encyclopedia of Alabama* (http://www.encyclopediaofalabama.org).

Oracles and Predictive Dreams

On predictive dreams, see Richard Wiseman (2012), *Paranormality. The Science of the Supernatural,* London: Pan Books, pp. 271–309; For the statistical argument, see e.g. John Allen Paulos (1989), *Innumeracy. Mathematical Illiteracy and Its Consequences*, New York: Hill and Wang, p. 52. The example of the Enschede fireworks disaster is taken from Victor Spoormaker (2003), *Alles over dromen* (*Everything about Dreams*), pp. 138–139. About the disaster itself, see Wikipedia "Enschede fireworks disaster".

On Paul the Octopus, see David Spiegelhalter (2011), "The Maths of Paul the 'Psychic' Octopus" (https://understandinguncertainty.org/node/888). See also the Wikipedia article "Paul the Octopus". Many newspaper articles about the prophetic animal appeared at the time.

A Modern Miracle

On the infinitesimal chance of being dealt four perfect hands, see N.T. Gridgeman (1964), "The Mystery of the Missing Deal," *The American Statistician*, Vol. 18, No. 1 (Feb.), pp. 15–16. See also the newspaper articles of November 24, 2011, mentioned in the text.

The Birthday Paradox

An enlightening article is Steven Strogatz (2012), "It's My Birthday Too, Yeah," *The New York Times Opinionator blog* (http://opinionator.blogs.nytimes.com/2012/10/01/its-my-birthday-too-yeah). For a more mathematical treatment see the chapter "Coinciding Birthdays" in Gorroochurn (2012), pp. 240–246.

The excerpt from the *Tonight Show with Johnny Carson* can be found on www.cornell.edu/video/the-tonight-show-with-johnny-carson-feb-6-1980-excerpt.

On the randomness of the iPod Shuffle function, see "Do iPods really shuffle?" on www.mathscareers.org.uk/article/ipods-really-shuffle.

Lifting the Fog

A biography of Bayes is David R. Bellhouse (2004), "The Reverend Thomas Bayes, FRS: A Biography to Celebrate the Tercentenary of His Birth," *Statistical Science* 19, no. 1, pp. 3–43.

Bayes' rule is explained on https://www.mathsisfun.com/data/bayes-theorem.html.

For Eddy's study, see David M. Eddy, "Probabilistic reasoning in clinical medicine: problems and opportunities," in: Daniel Kahneman, Paul Slovic and Amos Tversky (eds.), *Judgment under Uncertainty:*

Heuristics and Biases, Cambridge, U.K.: Cambridge University Press, pp. 249–267.

There exists a vast literature on math anxiety. A good introduction from a practical point of view is Christie Blazer (2011), "Strategies for Reducing Math Anxiety," *Information Capsule*, vol. 1102 (available from https://eric.ed.gov/?id=ED536509). Also DesLey V. Plaisance (2017), "A Teacher's Quick Guide to Understanding Mathematics Anxiety," *LATM Journal*, vol. 6, no. 1 (available from https://lamath.org/journal/). A pioneering study from an academic perspective is Sheila Tobias and Carol Weissbrod (1980), "Anxiety and Mathematics: An Update," *Harvard Educational Review*, vol. 50, no. 1, pp. 63–70. See also the article "Mathematical anxiety" on Wikipedia.

An account of Gigerenzer's interview of the 48 German doctors can be found in Gerd Gigerenzer (1996), "The Psychology of Good Judgement: Frequency Formats and simple algorithms," *Medical Decision Making* 16, pp. 273–280, and also in Gerd Gigerenzer (2002), *Calculated Risks*, New York: Simon & Schuster), pp. 39–86 (= Gerd Gigerenzer, *Reckoning with Risk*, London: Penguin Books Ltd). Steven Strogatz (2014), *The Joy of X*, London: Atlantic Books, pp. 183–190 also discusses the interview.

On the frequency format, see Gerd Gigerenzer and Ulrich Hoffrage (1995), "How to improve Bayesian reasoning without instruction: Frequency formats," *Psychological Review* 102, pp. 684–704.

The False Positive Paradox

On the false positive paradox, see "False Positives and False Negatives" on
https://www.mathsisfun.com/data/probability-false-negatives-positives.html.

Covid-19

On Covid-19 tests and the problem of false positives, see Christian Yates (2020), "Coronavirus: surprisingly big problems caused by small errors in testing", *The Conversation*, May 5 (https://theconversation.com/coronavirus-surprisingly-big-problems-caused-by-small-errors-in-testing-136700), and Priyanka Gogna (2020), "What do you mean, it was a false positive? Making sense of and terminology," *The Conversation*, September 8, (https://theconversation.com/what-do-you-mean-it-was-a-false-positive-making-sense-of-covid-19-tests-and-terminology-144484).

The percentages of infected people in New York City and London are taken from respectively (https://www.amny.com/coronavirus/almost-a-quarter-of-nyc-residents-test-positive-for-covid-19-antibodies-cuomo/) and (https://www.cityam.com/ten-per-cent-of-londons-population-may-have-caught-coronavirus/).

On the "immunity passport", see e.g. David Hagmann, George Loewenstein and Peter A. Ubel (2020), "Antibody tests might be deceptively dangerous. Blame the math," *The Washington Post*, April 30.

The Trial of the Century

On the fallacy committed by the defense in the O.J. Simpson case, see Irving J. Good (1995), "When Batterer Turns Murderer," *Nature* 375, p. 541, and Irving J. Good (1996), "When Batterer Becomes Murderer," *Nature* 381, p. 481 (both articles are downloadable from www.nature.com).

Good made use of Bayes' rule. For an approach using frequencies, see Steven Strogatz (2014), *The Joy of X*, London: Atlantic Books, pp. 183–190 (published earlier in the *New York Times*, "Chances are", April 25, 2010).

About the murder case itself, see the English Wikipedia article "O. J. Simpson Murder Case" and, from the perspective of the defense, Alan M. Dershowitz (1997), *Reasonable Doubts: The Criminal Justice System and the O.J. Simpson Case*, New York: Touchstone Books.

J'Accuse…!

The original French report by Henri Poincaré and his two colleagues Gaston Darboux and Paul Appell is available from http://henripoincarepapers.univ-lorraine.fr/chp/hp-pdf/hp1909ad.pdf. An abridged English translation as well as some secondary literature relating to the report can be found on https://www.maths.ed.ac.uk/~v1ranick/dreyfus.htm.

On the Dreyfus Affair and the misuse of mathematics, see David H. Kaye (2007), "Revisiting 'Dreyfus': A More Complete Account of a Trial by Mathematics", *Minnesota Law Review*, Vol. 91, No. 3, pp. 825–835, and Leila Schneps and Coralie Colmez (2013), *Math on Trial. How Numbers Get Used and Abused in the Courtroom*, "Math Error Number 10", New York: Basic Books.

The famous newspaper-article by Emile Zola can be found on Wikisource (https://fr.wikisource.org/wiki/J'accuse...!). An English translation is also available from Wikisource (https://en.wikisource.org/wiki/Translation:J'Accuse...!).

On the affair, see the English and French Wikipedia articles on the Dreyfus Affair.

An Angel of Death

The term *prosecutor's fallacy* was coined by William C. Thompson and Edward L. Schumann (1987), "Interpretation of statistical evidence in criminal trials — The Prosecutor's Fallacy and the Defense Attorney's Fallacy", *Law and Human Behavior* 11 (3), pp. 167–187.

On the Sally Clarke case, see Leila Schneps and Coralie Colmez (2013), *Math on trial*, "Math error number 1".

On the abuse of mathematics in the Lucia de Berk case and, see Leila Schneps and Coralie Colmez (2013), *Math on trial*, "Math error number 7"; Ben Goldacre (2007), "Lies, damned lies and statistics" on https://www.badscience.net/2007/04/losing-the-lottery/#more-392; and Mark Buchanan (2007) "Statistics: Conviction by numbers," *Nature* 445, pp. 254–255.

For a critical evaluation of the flawed model used by the expert witness, see Michiel van Lambalgen, Ronald Meester, Richard Gill and Marieke Collins (2007), "On the (ab)use of statistics in the legal case against the nurse Lucia de B.," *Law, Probability and Risk*, vol. 5, pp. 233–250, and Richard D. Gill, Piet Groeneboom, Peter de Jong

(2010), "Elementary Statistics on Trial (the case of Lucia de Berk)" available from http://www.math.leidenuniv.nl/~gill/hetero6.pdf.

Detailed information can be found on the website of the Lucia de Berk committee (http://www.luciadeb.nl/english). For a general overview, see the English Wikipedia-article on Lucia de Berk.

The Monty Hall Dilemma

Marilyn vos Savant dedicated four of her columns to the Monty Hall Dilemma. They were published in *Parade Magazine* on September 9, 1990; December 2, 1990; February 17, 1991; and July 7, 1991.

For the original question from a reader and Marilyn's answer, as well as an excerpt from the countless reactions she received, see Marilyn vos Savant (1996), *The Power of Logical Thinking*. New York: St. Martin's Press, pp. 5–16; see also the first three chapters of Donald Granberg (2014), *The Monty Hall Dilemma: A Cognitive Illusion Par Excellence*. Salt Lake City: Lumad Press. See also Donald Granberg, "The Monty Hall Dilemma. To Switch or not to Switch" in the appendix of Marilyn vos Savant (1996). Further information is provided on Marilyn's website (http://marilynvossavant.com).

The front-page article of *The New York Times* is "Behind Monty Hall's Doors: Puzzle, Debate and Answer?" by John Tierney, published on July 21, 1991.

The letters Steve Selvin dedicated to the Monty Hall dilemma are: Steve Selvin (1975a), "A problem in probability (letter to the editor)", *American Statistician*, Vol. 29, No. 1 (February 1975), p. 67, and: Steve Selvin (1975b), "On the Monty Hall problem (letter to the editor)", *American Statistician*, Vol. 29, No. 3 (August 1975), p. 134. Both letters are also included as appendix B in Granberg (2014).

A monography on the Monty Hall dilemma is Jason Rosenhouse (2009), *The Monty Hall Problem. The Remarkable Story of Math's Most Contentious Brain Teaser*, New York: Oxford University Press.

On Paul Erdös and the Monty Hall dilemma, see Andrew Vazsonyi (2002), *Which Door has the Cadillac. Adventures of a Real-Life Mathematician*, Lincoln: iUniverse, Inc., pp. 1–9, and the acclaimed biography by Paul Hoffman (1998), *The Man Who Loved Only Numbers*, London: Fourth Estate Limited, pp. 233–240.

Further literature can be found on Wikipedia. Google will direct you to countless websites on the Monty Hall dilemma, with or without simulation.

Ten Challenging Problems

A classic is Frederick Mosteller (1987), *Fifty Challenging Problems in Probability with Solutions*, New York: Dover Publications Inc.

Monty Hall 2.0 was published in the *Guardian* on August 24, 2020 by Alex Bellos, "Can you solve it? Win the car, dodge the goat", and is available on https://www.theguardian.com/science/2020/aug/24/can-you-solve-it-win-the-car-dodge-the-goat.

List of Illustrations in Part One

Bernoulli's Ars Conjectandi — Title page from *Ars Conjectandi*, Basel 1713.

The Bernoulli brothers — Engraving from Louis Figuier, *Vie des savants illustres du XVIIIe siècle*, Paris 1870.

Index

Printed in the United States
by Baker & Taylor Publisher Services